CLIMATE CHANGE AND ITS CAUSES, EFFECTS AND PREDICTION

CLIMATE CHANGE

EXTREME WEATHER, RISKS AND COSTS

Climate Change and its Causes, Effects and Prediction

Additional books and e-books in this series can be found on Nova's website under the Series tab.

CLIMATE CHANGE AND ITS CAUSES, EFFECTS AND PREDICTION

CLIMATE CHANGE

EXTREME WEATHER, RISKS AND COSTS

NICOLAS DELGADO
EDITOR

snova
New York

Library of Congress Cataloging-in-Publication Data

ISBN: 978-1-53619-113-4

Published by Nova Science Publishers, Inc. † New York

CONTENTS

PREFACE

Chapter 1 examines the state of the science related to extreme weather events. It will provide an opportunity to examine the role of climate change and other weather and climate factors in causing and exacerbating extreme weather events, to discuss economic and other societal impacts of extreme weather, to explore the state of forecasting and prediction of extreme weather with a focus on how to communicate uncertainty, and to identify gaps in the science.

Chapter 2 reports on the kinds of climate risk information standards and tools that communities need to reduce the risks and costs of climate change, including more extreme floods and wildfires.

In: Climate Change
Editor: Nicolas Delgado
ISBN: 978-1-53619-113-4

Chapter

UNDERSTANDING, FORECASTING, AND COMMUNICATING EXTREME WEATHER IN A CHANGING CLIMATE*

Committee on Science, Space, and Technology

Thursday, September 26, 2019
House of Representatives,
Committee on Science, Space, and Technology,
Washington, D.C.

The Committee met, pursuant to notice, at 10:01 a.m., in room 2318 of the Rayburn House Office Building, Hon. Eddie Bernice Johnson [Chairwoman of the Committee] presiding.

* This is an edited, reformatted and augmented version of Hearing before the Committee on Science, Space, and Technology, House of Representatives, One Hundred Sixteenth Congress, First Session, Serial No. 116–49, dated September 26, 2019.

Committee on Science, Space, and Technology U.S. House of Representatives, Hearing Charter, Understanding, Forecasting, and Communicating Extreme Weather in a Changing Climate

Thursday, September 26, 2019
10:00 a.m.

Purpose

The purpose of this hearing is to understand the state of the science related to extreme weather events. This hearing will provide an opportunity to examine the role of climate change and other weather and climate factors in causing and exacerbating extreme weather events, to discuss economic and other societal impacts of extreme weather, to explore the state of forecasting and prediction of extreme weather with a focus on how to communicate uncertainty, and to identify gaps in the science.

Witnesses

- Dr. J. Marshall Shepherd (SHEP-erd), Georgia Athletic Association Distinguished Professor of Atmospheric Sciences and Geography, Director, Atmospheric Sciences Program, Department of Geography, University of Georgia, 2013 President, American Meteorological Society
- Dr. James Done (DOE-n), Project Scientist III and Willis Research Fellow, Capacity Center for Climate & Weather Extremes, Mesoscale & Microscale Meteorology Lab, National Center for Atmospheric Research
- Dr. Adam Sobel (SO-Bull), Professor, Lamont-Doherty Earth Observatory and School of Engineering and Applied Sciences,

Columbia University, Director and Chief Scientist, Initiative on Extreme Weather and Climate, Columbia University

- Dr. Berrien Moore (MORE), Director, National Weather Center, University of Oklahoma
- Dr. Ann Bostrom (BOS-trum), Weyerhaeuser Endowed Professor in Environmental Policy, University of Washington

Overarching Questions

- What is the current state of science on understanding the causes of extreme weather events?
- What are the greatest challenges associated with forecasting extreme weather events, and how do we improve forecasts and models?
- What is the relationship between climate change and extreme weather events?
- What impacts does extreme weather supercharged by climate change have on vulnerable people and property?
- How can we improve the way extreme weather forecasts are communicated to the public to ensure people take the necessary safety precautions?
- What are the current federal funding needs for enhanced observations, model improvements, and research and development activities related to extreme weather?

Background

July 2019 was the hottest month on record for the planet. Temperatures soared worldwide, in a global pattern of warming that has increased in recent years.[1] Though the United States is no stranger to extreme weather

---

[1] https://www.noaa.gov/news/july-2019-was-hottest-month-on-record-for-planet.

events, with different parts of the country experiencing heatwaves, cold snaps, wildfires, tornadoes, hurricanes, and flooding, there has been an increase in the frequency and intensity of certain extreme weather events over the past few years.[2]

The fifth IPCC Assessment Report (AR5), details the observed changes in extreme weather events that have occurred since 1950.[3] It notes that decreases in cold temperature extremes, increases in warm temperature extremes, increases in heavy precipitation events in certain regions, and an increase in extreme high sea level, can be firmly linked to human influences.[4] It also concludes that "[i]mpacts from recent climate-related extremes, such as heat waves, droughts, floods, cyclones and wildfires, reveal significant vulnerability and exposure of some ecosystems and many human systems to current climate variability."[5]

The ability to accurately predict extreme weather has advanced greatly in the last few decades; however, there is much room for improvement.[6] Recent extreme weather events such as Hurricane Dorian have highlighted the uncertainty that remains in understanding and predicting severe storms, as well as the need for further investment in observations, improvements to models, and overall research and development activities. Accurate forecasts are instrumental in preparing our communities for extreme weather events and minimizing property damage and injury. Investigating the underlying causes of extreme events and how they have changed over time helps to improve forecasting capabilities.[7]

Communicating extreme weather forecasts is an ongoing issue for local, state, and federal officials. More social and behavioral science research is needed to understand how people interpret and respond to

---

[2] Third National Climate Assessment. 2014. https://nca2014.globalchange.gov/highlights/report-findings/extreme-weather.

[3] https://www.ipcc.ch/site/assets/uploads/2018/02/AR5_SYR_FINAL_SPM.pdf.

[4] Ibid.

[5] Ibid.

[6] National Academies of Sciences, Engineering, and Medicine. 2016. *Attribution of Extreme Weather Events in the Context of Climate Change*. Washington, DC: The National Academies Press. https://doi.org/10.17226/21852.

[7] Ibid.

weather forecasts.[8] Even with improvements in prediction and warning systems, extreme events still result in great loss of life and other preventable costs. Impacts of weather are strongly dependent on behavioral responses to forecasts; thus, increased investment in social science is needed to understand how social factors affect how the public prepares for and responds to extreme weather events.[9]

Certain populations are more vulnerable to extreme weather than others. The poor, elderly, disabled, people of color, and other at-risk groups are more likely to experience negative health and other impacts of severe weather.[10] There is an urgent need to understand the intersection between extreme weather and vulnerable communities in order to prepare and protect them from increasingly damaging and frequent extreme weather events.

Understanding the Causes of Extreme Weather Events

The underlying physical processes that lead to the development of extreme weather events are in different stages of understanding. For example, scientists have a high level of understanding of how thunderstorms, hurricanes, and droughts form.[11] There remains a need for additional research to better understand the fundamental physical processes behind the formation of other types of severe weather, such as tornadoes, which less well understood.[12]

The nascent field of attribution science seeks to determine if extreme weather events can be attributed to anthropogenic climate change.[13] Some events are easier than others to confidently attribute to climate change. Confidence is strongest for drought and heavy precipitation events, and

---

[8] National Academies of Sciences, Engineering, and Medicine. 2018. *Integrating Social and Behavioral Sciences within the Weather Enterprise*. Washington, DC: The National Academies Press. https://doi.org/10.17226/24865.

[9] Ibid.

[10] https://nca2018.globalchange.gov/downloads/NCA4_2018_FullReport.pdf.

[11] https://www.nssl.noaa.gov/education/svrwx101/.

[12] https://www.nssl.noaa.gov/education/svrwx101/tornadoes/.

[13] NASEM. 2016. https://doi.org/10.17226/21852.

events related to temperature, such as extreme heat or cold.[14] For these events, there is a well understood and simulated physical mechanism that relates it to long-term human-caused climate change. Events that are more difficult to attribute to climate change may be complicated by natural variability or a lack of understanding of the basic underlying science, such as tornadoes. In addition, non-meteorological factors can limit the accuracy of model simulations of some events, such as with drought or wildfire.

A continued, focused effort to improve specific aspects of the analysis of weather and climate extremes is needed to improve both attribution science and the underlying physics of severe weather.[15] Many attribution analyses depend on estimations of event probabilities or distributions of event magnitude simulated by models. Consequently, confidence in attribution results relies on the skill of the model. More emphasis on evaluating models is needed across studies on event attribution.

Forecasting Extreme Weather Events

Developments in observations, numerical monitoring, and data assimilation have led to major improvements in weather forecasts over the past few decades. At present, a 5-day forecast is as accurate as a 1-day forecast in 1980.[16] Global data collection systems contribute daily to forecasting models. There are more than 10,000 manned and automatic surface weather stations, 1,000 upper-air stations, 7,000 ships, more than 1,000 buoys, 3,000 aircraft, a constellation of Earth-observing satellites, and hundreds of radars that monitor and collect land, atmosphere, and ocean condition data every day.[17]

Enhancing our understanding of the underlying science of extreme weather, coupled with sustained observations and improvements in modeling, are needed to increase forecasting capabilities of these events. Forecasting goals include the ability to deliver predictive forecasts of

[14] Ibid.
[15] Ibid.
[16] https://science.sciencemag.org/node/721705.full.
[17] https://public.wmo.int/en/our-mandate/what-we-do/observations.

future extreme weather events with lead times of days, seasons, or longer, and aim to account for natural variability and anthropogenic influences on the weather-climate system.[18] In addition, the Earth-observing satellite system must be protected from interference by 5G operations to safeguard future capabilities. Additional information on National Oceanic and Atmospheric Administration (NOAA) forecasting and potential impacts to forecasting due to 5G spectrum auctions can be found in the Charter from the May 16, 2019 Subcommittee on Environment Hearing titled *The Future of Forecasting: Building a Stronger U.S. Weather Enterprise.*[19]

Communicating Extreme Weather Events

Even though reliable forecasts of some extreme weather events are available days in advance, severe weather continues to cause avoidable deaths and other losses. There is a "growing recognition that a host of social and behavioral factors affect how we prepare for, observe, predict, respond to, and are impacted" by extreme weather.[20] In order to receive the greatest return on meteorological research and numerical weather prediction, the social and behavioral sciences (SBS) must be fully engaged in the communication of extreme weather.

There are a multitude of factors that affect the way in which people interpret forecasts. The effect of past weather experiences, and people's perceptions and attitudes, impact their response to weather forecasts and warning messages.[21] The effect of these messages depends on the characteristics of the audience receiving them. Questions remain as to how the weather enterprise can provide information when and where it is needed, and in the most accessible format for all stakeholders.[22]

Researchers are beginning to look into the design, interpretation, and effect of weather forecasts and messages on different populations. For

---

[18] NASEM. 2016. https://doi.org/10.17226/21852.
[19] https://docs.house.gov/Committee/Calendar/ByEvent.aspx?EventID=109467.
[20] NASEM. 2018. https://doi.org/10.17226/24865.
[21] Ibid.
[22] Ibid.

example, they have used eye-tracking technology to ascertain how people interpret weather messaging.[23] This type of technology can also provide insight into how forecasters make decisions about how to use information in complex forecasting situations.[24] Gaps remain in message design research, and more studies are needed to determine which types of messages have the maximum impact. In addition, more focus is needed on how forecasters can effectively communicate uncertainty in forecasts.

There is an increasing emphasis within the weather enterprise that SBS must be integrated into research and operations to improve public safety in the face of extreme weather.[25] To accomplish this, there must be investment in leadership to build awareness, and capacity built throughout the weather enterprise, with a focus on critical knowledge gaps in order to integrate SBS knowledge into weather forecasting and communications. Over the past year, some steps have been taken within the weather enterprise, but progress remains slow.[26] More emphasis on building SBS capacity is needed from the research to the operational stage, at every level of government and in partnership with the private sector and academia. The federal government can foster collaborations with social scientists, institutions of higher education, and private sector actors. Platforms should be created for interagency and intersectoral efforts, such as workshops, interagency working groups, university-based centers with federal funding, or other intersectoral mechanisms for achieving SBS integration into the weather enterprise.[27]

Impacts of Extreme Weather Events

Many sectors of U.S. society are vulnerable to extreme weather events and their resulting costs.[28]

---

[23] Ibid.
[24] Ibid.
[25] Ibid.
[26] Ibid.
[27] Ibid.
[28] Fourth National Climate Assessment. 2018. https://nca2018.globalchange.gov/downloads/NCA4_2018_FullReport.pdf.

A reliable, safe and efficient U.S. transportation system is threatened by heavy precipitation, coastal flooding, heat, wildfires, and other extreme events. Extreme heat events can have a negative impact on pavement and flooding will cause vehicle delays; both events will increase road maintenance costs and disproportionately impact vulnerable populations. Coastal communities will suffer from the increasing frequency of flooding events and sea level rise. By the middle of the century, flooding and storms are likely to destroy billions of dollars of property, with the Atlantic and Gulf regions facing the greatest risk. Forest structure and function will change rapidly with increasingly severe ecological disturbances caused by extreme weather. Agricultural and rural communities will face degradation of soil and water resources with an increase in extreme precipitation events.[29]

A recent study found that in 2012, 10 extreme weather events, boosted by climate change, caused $10 billion to the U.S. healthcare system.[30] The study looked at 10 climate-sensitive events across 11 states, including Superstorm Sandy, wildfires in Colorado and Washington, extreme heat in Wisconsin, harmful algal blooms in Florida, and more. These events led to 1,000 more deaths and over 20,000 additional hospitalizations. Two-thirds of these public health costs were paid by Medicare and Medicaid.[31]

The study is the first of its kind to investigate the economic toll of climate-sensitive public health costs of extreme weather. Additional research is needed to more fully understand the financial toll of deaths, hospitalizations, visits to the emergency room, and associated medical care as a result of extreme weather events. In order to reduce the health-related costs of extreme weather, improvements must be made in public health preparedness, resource deployment, and public outreach and communication, particularly in vulnerable communities.[32]

---

[29] Ibid.

[30] Estimating the Health-Related Costs of 10 Climate-Sensitive U.S. Events during 2012. 2019. https://doi.org/10.1029/2019GH000202.

[31] Ibid.

[32] Ibid.

As of July 9, 2019, the U.S. has experienced six weather and climate disasters that had losses exceeding $1 billion each.[33] Without mitigation and adaptation, these numbers will continue to grow as more extreme events occur. Extreme weather events will also affect other countries, and can slow or reverse social and economic progress in developing countries, exacerbate conflict, and negatively impact global trade. This poses a major threat to U.S. national security, as does the risks from extreme weather impacts on military assets, such as roads, runways, and other infrastructure, both at home and abroad.[34]

According to the 2018 Intergovernmental Panel on Climate Change (IPCC) Special Report on 1.5°C, every increment of warming will exacerbate the impacts of extreme weather worldwide. For example, more people will experience extreme heat and drought, particularly in urban areas.[35] In high latitude and mountainous regions, there will be heavier precipitation and flooding.[36]

For information on the Special Report on Global Warming of 1.5°C and its findings on the relationship between climate change and extreme weather, see the Charter from the February 13 Hearing titled *The State of Climate Science and Why it Matters*.[37]

Additional Reading

The 1.5 Health Report: Synthesis on Health & Climate Science in the IPCC SR1.5. https://www.who.int/globalchange/181008_the_1_5_healthreport.pdf.

United in Science: High-level synthesis report of latest climate science information convened by the Science Advisory Group of the UN Climate Action Summit 2019. https://public.wmo.int/en/resources/united_in_science.

[33] https://www.ncdc.noaa.gov/billions/.
[34] https://nca2018.globalchange.gov/downloads/NCA4_Report-in-Brief.pdf.
[35] https://report.ipcc.ch/sr15/pdf/sr15_chapter3.pdf.
[36] Ibid.
[37] https://docs.house.gov/Committee/Calendar/ByEvent.aspx?EventID=108915.

Weather, Climate & Catastrophe Insight, 2018 *Annual Report*. http://thoughtleadership.aonbenfield.com/Documents/20180124-ab-if-annual-report-weather-climate-2017.pdf.

* * *

Chairwoman JOHNSON. This hearing will come to order. Without objection, the Chair is authorized to declare a recess at any time.

Good morning, and welcome to our day's hearing on extreme weather. This is a topic that I think is universally relevant, as many of my colleagues and our constituents have dealt with extreme weather events recently. In fact, NOAA's (National Oceanic and Atmospheric Administration's) National Centers for Environmental Information found that in 2018 alone the U.S. experienced 14 climate weather and disasters with losses for each topping $1 billion. These events included drought, severe storms, wildfires, tropical cyclones, and winter storms, and they impacted nearly every State in the continental U.S.

As of July 2019, the U.S. has already experienced six weather and climate events with losses greater than $1 billion dollars each. And July 2019 was also the hottest month on record worldwide, which led to record low levels of sea ice in both the Arctic and Antarctic.

There is an increasing scientific consensus that human-driven climate change is playing an undeniable role in many of the extreme weather events that we have experienced. Earlier this week, the World Meteorological Organization (WMO) of the U.N. released a report that found that climate change, through the slowing of the jet stream, could be dircctly linked to the record-breaking heatwaves experienced across North America, Europe, and Africa in 2018 and 2019. There was also clear evidence that the jet stream pattern influenced many extreme rainfall events as well.

Yesterday, thc IPCC (Intergovernmental Panel on Climate Change) released a special report on the oceans and cryosphere. It identified up to 90 percent of marine heatwaves from 2006 to 2015 were due to climate change. Climate change was also responsible for the increased precipitation, winds, and extreme sea level events associated with some tropical cyclones. The special report also determined that some back-to-back extreme weather events that we have become accustomed to seeing have also been influenced by climate change.

I know many of my colleagues from the Houston Gulf Coast area have directly experienced these impacts with the extreme rainfall that they saw from Hurricane Harvey 2 years ago. And most recently, they had to deal with Tropical Storm Imelda, which dropped over 40 inches of rain in some parts of Houston just last week.

This hearing is especially timely given not only recent extreme weather events such as Dorian and Imelda, but also because September is National Preparedness Month. It is important for our constituents to understand how they can and should be preparing for disasters, including extreme weather. It is vitally important that the public can rely on official forecasts from the National Weather Service to inform their responses to weather events without worrying that these forecasts have been interfered with.

Though our ability to forecast the path of a storm like Hurricane Dorian has greatly improved, our dedicated meteorologists in the National Weather Service still cannot say with absolute certainty what the intensity of a storm like this will be. These track forecasts have relied heavily on satellite observations, and any interference with the data in these observations, such as water vapor measures, could have dire consequences for communities that lie in the path of a similar hurricane.

We have discussed in this Committee the importance of sustained observations to feed into the weather models that are used to develop forecasts and the need to continually be improving those models and subsequent forecasts. I expect today's hearing will be no different.

But, in addition to the need to continue to support the physical science, observations, and modeling that goes into developing forecasts, there is also a need to understand how to better integrate the social and behavioral sciences in our weather enterprise. More research is needed to understand how our biases can impact the forecasting process, and how our past experiences with extreme weather events can influence how the public interprets forecasts and notices from emergency managers.

With that, I would like to extend my welcome to our very distinguished panel and thank them for joining us this morning. We are looking forward to a robust discussion on how this Committee can help our country better prepare for future extreme weather events, events that we are likely to expect more frequently and intensity due to climate change.

[The prepared statement of Chairwoman Johnson follows:]

CHAIRWOMAN EDDIE BERNICE JOHNSON (D-TX), FULL COMMITTEE HEARING: UNDERSTANDING, FORECASTING, AND COMMUNICATING EXTREME WEATHER IN A CHANGING CLIMATE, THURSDAY, SEPTEMBER 26, 2019

Good morning and welcome to today's hearing on extreme weather. This is a topic that I think is universally relevant, as many of my colleagues and our constituents have dealt with extreme weather events recently.

In fact, NOAA's National Centers for Environmental Information found that in 2018 alone the U.S. experienced 14 climate weather and disasters with losses for each topping $1 billion dollars. These events included drought, severe storms, wildfires, tropical cyclones, and winter storms, and they impacted nearly every state in the continental U.S. As of July 2019, the U.S. has already experienced six weather and climate events with losses greater than $1 billion dollars each. July 2019 was also the hottest month on record worldwide, which led to record low levels of sea ice in both the Arctic and Antarctic.

There is an increasing scientific consensus that human-driven climate change is playing an undeniable role in many of the extreme weather events that we have experienced.

Earlier this week the World Meteorological Organization of the UN released a report that found that climate change, through the slowing of the jet stream, could be directly linked to the record-breaking heatwaves experienced across North America, Europe, and Africa in 2018 and 2019. There was also clear evidence that this jet-stream pattern influenced many extreme rainfall events as well.

Yesterday the IPCC released a Special Report on the oceans and cryosphere. It identified that up to 90% of marine heatwaves from 2006 to 2015 were due to climate change. Climate change was also responsible for the increased precipitation, winds, and extreme sea level events associated with some tropical cyclones. The Special Report also determined that some back to back extreme weather events that we have become accustomed to seeing have also been influenced by climate change.

I know many of my colleagues from the Houston and Gulf Coast area have directly experienced these impacts with the extreme rainfall that they saw from Hurricane Harvey two years ago. And most recently they had to deal with Tropical Storm Imelda, which dropped over 40 inches of rain in some parts of Houston just last week.

This hearing is especially timely given not only recent extreme weather events such as Dorian and Imelda, but also because September is National Preparedness Month. It is important for our constituents to understand how they can and should be preparing for disasters, including extreme weather. It is vitally important that the public can rely on official forecasts from the National Weather Service to inform their response to weather events without worrying that these forecasts have been interfered with.

Though our ability to forecast the path of a storm like Hurricane Dorian has greatly improved, our dedicated meteorologists in the National Weather Service still cannot say with absolute certainty what the intensity of a storm like that will be. These track forecasts have relied heavily on satellite observations, and any interference with the data in these observations, such as water vapor measurements, could have dire consequences for communities that lie in the path of similar hurricanes.

We have discussed in this Committee the importance of sustained observations to feed into the weather models that are used to develop forecasts, and the need to continually be improving those models and subsequent forecasts. I expect today's hearing will be no different. But, in addition to the need to continue to support the physical science, observations, and modeling that goes into developing forecasts, there is also a need to understand how to better integrate the social and behavioral sciences in our weather enterprise. More research is needed to understand how our biases can impact the forecasting process, and how our past experiences with extreme weather events can influence how the public interprets forecasts and notices from emergency managers.

With that, I would like to extend my welcome to our very distinguished panel and thank them for joining us this morning. We are looking forward to a robust discussion on how this Committee can help our country better prepare for future extreme weather events; events that we are likely to expect with more frequency and intensity due to climate change.

Thank you.

* * *

Chairwoman JOHNSON. At this time I recognize our Ranking Member, Mr. Lucas, for an opening statement.

Mr. LUCAS. Thank you, Chairwoman Johnson, for holding today's hearing on extreme weather, an important topic within our Committee's jurisdiction.

Extreme weather events are of concern to us all, regardless of what part of the country we represent. These events represent threats to lives, property and often occur with little warning. The economic toll of extreme weather events across the Nation is significant. The most recent U.S. National Climate Assessment (USNCA) stated 241 incidents with more than $1 billion of economic damage since 1980, including 14 such events in 2018 alone.

I want to make my position clear. The climate is changing; global industrial activity has played a role in this. The complex relationship between climate and weather is in need of continued research, and this Committee has a responsibility to prioritize that research if we want America to be a global leader in this field. That should be the focus of today's hearing, not outright denial or finger-pointing on inaction. There are many components as we examine how best to research and respond to extreme weather, including how we help communities prepare for these events, how we improve our weather forecasts, and how we communicate the possibility of an extreme weather event to our citizens.

This Committee has taken steps to help address these issues. The *Weather Act*, signed into law in 2017, directed NOAA to improve its tornado warning capabilities and hurricane forecasting capacity, two of the most destructive types of extreme weather events. Additionally, the legislation required NOAA to perform an assessment of its practices on communicating extreme weather events to the public. NOAA has made progress in implementing these provisions in the last 2 years, but much work remains.

A *Weather Act* reauthorization was signed into law in January. Included in the legislation was congressional authority of NOAA's Earth Prediction Innovation Center, commonly known as EPIC. This initiative will make the National Weather Service's numerical prediction models available to the academic community for crowdsource forecasting on a larger scale, which in turn will help improve our national forecasting ability.

Oklahoma is no stranger to extreme weather events. Whether it's an outbreak of tornadoes, severe droughts affecting our farmers and ranchers, or extended cold weather, we have seen it all. Thankfully, Oklahoma is home to one of the world's most renowned experts in the field of weather research and forecasting. The National Weather Center is located on the University of Oklahoma campus in Norman and houses Federal, State, and university researchers in a collaborative environment. Among the Federal offices in the Weather Center are NOAA's National Severe Storms Laboratory and Storm Prediction Centers. These offices are at the leading edge of researching and forecasting the outbreak of extreme weather events across the country.

Additionally, Oklahoma is home to the Nation's premier weather mesonet. A mesonet is a series of small weather stations spread out across a large area which help monitor real-time conditions on the ground and provide citizens and forecasters with vital data. This data helps farmers determine the optimal time to plant and can alert emergency responders if conditions are ripe for developing a tornado. As this Committee considers possible legislative initiatives based on today's hearing, we should look at the Oklahoma mesonet as a model for how we can improve the forecasting and communication of severe weather events.

Our panel of witnesses today bring diverse perspectives on researching all aspects of extreme weather events. I thank them for taking time to be here, and I look forward to a productive conversation on this important topic.

And with that, I yield back, Madam Chair.

[The prepared statement of Mr. Lucas follows:]

Opening Statement of Ranking Member Frank Lucas, Full Committee Hearing, Understanding, Forecasting, and Communicating Extreme Weather in a Changing Climate, September 26, 2019

Thank you, Chairwoman Johnson, for holding today's hearing on extreme weather and how we can better forecast and respond to it.

Extreme weather events are of concern to us all, regardless of which part of the country we represent. These events represent threats to lives and property and often occur with little warning. They also take a significant toll on our economy. The most recent National Climate Assessment cited 241 incidents with more than a billion dollars of economic damage since 1980, including 14 events in 2018 alone.

I want to make my position clear: the climate is changing, and global industrial activity has played a role in this. The complex relationship between climate and weather is in need of continued research. This Committee has a responsibility to prioritize that research so we can continue to mitigate storm damage, grow our economy, and provide certainty for businesses that depend on accurate forecasts.

This research should be the focus of today's hearing, because it's research that actually provides answers to the challenges we face.

As we examine how best to research and respond to extreme weather, there are a variety of factors to consider, including how we help communities prepare for these events, how we improve our weather forecasts, and how we communicate the possibility of an extreme weather event to our citizens.

This committee has taken steps to help address these issues. The Weather Act, signed into law in 2017, directed NOAA to improve its tornado warning capabilities and hurricane forecasting capacity – two of the most destructive types of extreme weather events. Additionally, the legislation required NOAA to perform an assessment of its practices on communicating extreme weather events to the public. NOAA has made progress in implementing these provisions in the last two years, but much work remains.

A Weather Act reauthorization was signed into law in January which authorized NOAA's Earth Prediction Innovation Center – more commonly known as EPIC. This initiative will make the National Weather Service's numerical prediction models available to the academic community to crowdsource forecasting on a larger scale – which in turn will help improve our national forecasting ability.

Oklahoma is no stranger to extreme weather events. Whether it is an outbreak of tornadoes, severe droughts affecting our farmers and ranchers, or extended cold weather – we have seen it all.

Thankfully, Oklahoma is home to some of the world's most renowned experts in the field of weather research and forecasting. The National Weather Center is located on the University of Oklahoma campus in Norman and houses federal, state, and university researchers in a collaborative environment. Among the federal offices in the Weather Center are NOAA's National Severe Storms Laboratory and Storm Prediction Center. These offices are at the leading edge of researching and forecasting the outbreak of extreme weather events across the country.

Additionally, Oklahoma is home to the nation's premier weather mesonet. A mesonet is a series of small weather stations spread across a large area which help monitor real-time conditions on the ground and provide citizens and forecasters with vital data. This data helps farmers determine the optimal time to plant and can alert emergency responders if conditions are ripe for a developing tornado. As this Committee considers possible legislative initiatives based on today's hearing, we should look to the Oklahoma mesonet as a model for how we can improve the forecasting and communication of severe weather events.

Our panel of witnesses today brings diverse perspectives on researching all aspects of extreme weather events. I thank them for taking the time to be here and look forward to a productive conversation on this important topic.

* * *

Chairwoman JOHNSON. Thank you very much.

If there are Members who wish to submit additional opening statements, your statements will be added to the record at this point.

At this time, I'd like to introduce our witnesses. Our first distinguished witness is Dr. J. Marshall Shepherd. Dr. Shepherd is a leading international expert in weather and climate and is the Georgia Athletic Association Distinguished Professor of Geography and Atmospheric Sciences at the University of Georgia. Dr. Shepherd was the 2013 President of the American Meteorological Society. He is also the host of the Weather Channel's award-winning show Weather Geeks and a contributor to *Forbes* magazine. He was the first African American to receive a Ph.D. from the Florida State University Department of Meteorology.

Our second witness, Dr. James Done, is a Senior Willis Fellow and Deputy Director of the Capacity Center for Climate and Weather Extremes at the National Center of Atmospheric Research (NCAR). He works with shareholders from the energy, water, and insurance sectors to understand future weather and climate extremes and their impacts. Dr. Done received his Ph.D. in meteorology from the University of Reading in the U.K.

Our third witness is Dr. Adam Sobel. Dr. Sobel is a Professor at Columbia University's Lamont-Doherty Earth Observatory and the Fu Foundation School of Engineering and Applied Sciences and leads the Columbia University Initiative on Extreme Weather and Climate. He is an atmospheric scientist who specializes in the dynamics of climate and weather, particularly in the tropics, on timescales of days to decades. Dr. Sobel earned his Ph.D. in meteorology from the Massachusetts Institute of Technology.

The Chair now recognizes our Ranking Member Lucas to introduce our fourth witness, Dr. Moore.

Mr. LUCAS. Thank you, Chairwoman Johnson.

And Dr. Berrien Moore is the Director of the National Weather Center and Dean of the College of Atmospheric and Geographic Sciences at the University of Oklahoma. Dr. Moore is an internationally recognized Earth scientist who's been honored by NASA (National Aeronautics and Space Administration), NOAA, and multiple international organizations. He received his bachelor of science and mathematics degree from the University of North Carolina and his Ph.D. in mathematics from the University of Virginia. He was the coordinating lead author for the final chapter of the Intergovernmental Panel on Climate Change, Third Assessment Report, as such, has been honored for contributing to the 2007 Nobel Peace Prize awarded to IPCC.

Back in March I had the opportunity to visit the National Weather Center while in Norman, and I was lucky to receive a tour from the Director himself. You might say I'm biased, but I believe these facilities and the researchers there are the best in the world. And this is a testament to Dr. Moore's continued dedication to make the United States the gold standard in weather prediction.

And in recognition of his lifelong work in weather science, he's been the recipient of numerous honors, including NASA's Distinguished Public

Service Award, NOAA's Administrator Recognition Award. He's also an elected fellow of the American Meteorological Society and the International Academy of Aeronautics.

Thank you for making the trip here today, Dr. Moore. Thank you, Chairwoman.

Chairwoman JOHNSON. Thank you very much.

Our final witness is Dr. Ann Bostrom, who's the Weyerhaeuser Endowed Professor of Environmental Policy at the Daniel J. Evans School of Public Policy and Governance at the University of Washington. She studies risk perception, communication, and decisionmaking under uncertainty in context of weather change, hurricanes, earthquakes, and tsunamis. Dr. Bostrom co-chaired the National Academies' study committee integrating social and behavioral sciences within the weather enterprise. She holds a Ph.D. in public policy analysis from the Carnegie Mellon University.

So thanks to all of you for being here. And as witnesses, you should know that each will have 5 minutes for your spoken testimony. Your written testimony will be included in the record for the hearing. When you have completed your spoken testimony, we will begin with questions, and each Member will have 5 minutes to question the panel.

We will start with Dr. Shepherd.

Testimony of Dr. J. Marshall Shepherd, Georgia Athletic Association Distinguished Professor of Atmospheric Sciences and Geography, and Director, Atmospheric Sciences Program, Department of Geography, University of Georgia, and 2013 President, American Meteorological Society

Dr. SHEPHERD. I would like to thank Chairwoman Johnson, Ranking Member Lucas, colleagues on the Committee, for this opportunity to share my thoughts on the contemporary extreme weather and its context within a changing climate.

In 2013 I sat before the Senate Environment and Public Works Committee on a similar topic, and there's nothing that I would change about what I say today from what I said then except to amplify the message.

NOAA recently updated what constitutes the 1,000-year flood in Texas because the rainfall is changing. This has implications for the National Flood Insurance Program and infrastructure design. Tropical storm Imelda and Hurricane Dorian joined Michael, Harvey, and Maria as extreme events that either rapidly intensified, stalled, or inundated regions. Was it caused by climate change has become a very popular question, but it's an ill-posed question. Extreme weather attribution must be carefully considered and framed without hype, speculation, and social media debates.

In 2016 I served as a co-author on a study by the National Academies of Sciences, Engineering, and Medicine on attribution of extreme weather events in the context of climate change. A key finding is that we are able, with some degree of attribution, to link climate change with some degree of confidence, moderate to high. The fingerprint of climate change is imprinted on the intensity or frequency of contemporary heatwaves, extreme rainfall, drought, and, to some degree, hurricanes. There is little to no confidence in attribution of tornadic storms at this time, but the research continues.

Let me stop right here and emphasize, yes, yes, yes, climate changes naturally and always has. It's often amusing when people remind me, a degreed climate scientist, of this fact at dinner or on social media. But it's not an either-or proposition. It's an "and" proposition. Grass grows naturally, but it grows very differently when we fertilize the soil.

In 2018 there were $39 billion-plus disasters. According to insurance broker Aon, insured dollars totaled $90 billion, which is the fourth-highest inflation-adjusted number of such events. And of those events, the United States had 16 of them.

We must message these events as kitchen-table issues and challenges to our water and food supply, public health infrastructure, and national security. It's not about polar bears in the year 2080. Hurricane Michael devastated my home State of Georgia. Hardworking Georgia farmers lost pecans, peanuts, and cotton. But guess what? Americans buy peanut butter and buy T-shirts. They felt the impact, too.

Now, I want to quote from a book that I read, Ecclesiastes 1:7, "All streams flow into the sea, yet the sea is never full. To the place the streams come from, there they return again." This text perfectly captures the water cycle that we all likely learned about in fourth grade. And we know that water is essential to life and doesn't understand the concept of liberal or conservative. Yet the water cycle is changing, more extreme downpours, melting snowpacks, and overwhelmed stormwater infrastructure.

So what do we do to move forward? We have to keep observational modeling capacity robust. Challenges with rainfall forecasts from Tropical Depression Imelda and intensity changes with Michael affirm the need for the EPIC framework that was mentioned. The National Center for Atmospheric Research and NOAA and other partners are moving out on EPIC, and I believe it's a positive step to ensure a nimble and responsive U.S. weather modeling capacity. To make our weather models the best in the world for forecasting the 0- to 14-day range and at seasonal timescales, we need the fastest supercomputers available to accommodate the emerging volume of observational data from NOAA, NASA, and global partners. NASA is implementing its decadal survey recommended by the National Academies. NOAA is funding underwater gliders at the University of Georgia to help with hurricane intensity forecasts.

But the best forecasts are bad forecasts if people don't interpret them, so we need continued and strong investments in social sciences, which we'll hear about later today. We need to understand risk. Increasingly, my focus is on risk because there are vulnerable populations in our society that because of social status, elevated health risks, et cetera, are more vulnerable to these events.

So with that in closing, I challenge the Committee and our country to stand forward on advancing the resources that we know we need to move extreme weather forecasting forward. Thank you for this time.

[The prepared statement of Dr. Shepherd follows:]

Written Testimony of Dr. J. Marshall Shepherd, Georgia Athletic Association Distinguished Professor of Atmospheric Sciences and Geography and Director, Atmospheric Sciences Program at the University of Georgia, Former President of the American Meteorological Society, before the Committee on Science, Space, and Technology of the United States House of Representatives, September 26, 2019

Clarity on Extreme Weather-Climate Change Attribution, Messaging, and Steps Forward Key Takeaway Points

Your constituents, my fellow citizens, and my two kids feel the impacts of extreme weather in their lives and "kitchen table" issues. It's not about polar bears or the year 2080.

- Attribution studies affirm that the influence of climate change is found in contemporary extreme weather events affecting the United States.
- Extreme weather events impact people, infrastructure and processes more than "average" events. Our agricultural, energy, water, transportation, public health, and national security institutions are at risk, and this ultimately affects American households and beyond. We must message the risks and impacts

with such context in mind rather than using distant and abstract concepts.
- Robust science and technology resources must be sustained at local to federal levels.
- We must keep a close eye on the well-being of vulnerable populations and disparities in climate resiliency.

Introduction

I would like to thank Chairwoman Eddie Bernice Johnson and her colleagues on the Science, Space, and Technology Committee for an opportunity to share my thoughts and expertise on contemporary extreme weather and its context within a changing climate. In 2013, I sat before the Senate Environment and Public Works Committee on a similar topic.[38] Two key messages that I delivered to the Committee at that time were:

- Climate change is increasing the probability of extreme events, and in some cases may be strengthening their intensity or increasing their frequency (i.e., we are loading the dice towards more Sandy or blizzard type storms)
- There is strong evidence that increases in some types of extremes are linked to human-induced climate change, notably extreme heat, coastal flooding, and heavy downpours.

For other types of extremes, such as tornadoes, current evidence is much more limited. Since 2013, there is nothing that I would change about those statements except amplifying and affirming them with greater vigor.

While in elementary school, I did a science project called “Can A 6th Grader Predict The Weather?” As a child, I was in awe of the weather but naively never understood that it could fundamentally alter the lives of families, disrupt our economy, take human lives, or threaten national

---

38 https://www.epw.senate.gov/public/index.cfm/2013/2/senate-briefing-on-the-latest-climate-science.

security. I was just a boy from Canton, Georgia making weather instruments from things around the house and charting local weather patterns. That passion for the weather set the course for my educational and professional career. However, the blinders of youthful naivety are gone. Extreme weather within a changing climate is one of the most pressing concerns of this generation and those to come. From my lens as a scientist, professor, and the former president of the American Meteorological Society (AMS), I would like to discuss extreme weather attribution, messaging, and pathways forward.

Contemporary Extreme Weather within the Context of Climate Change

Earlier this month, talented colleagues at the National Hurricane Center and National Weather Service monitored Hurricane Dorian. Dorian was a human tragedy for parts of the Bahamas because it stalled in that region as a major hurricane with powerful winds, storm surge, and multiple feet of rainfall.[39] Dorian eventually paralleled the U.S. coast as forecasted and created challenges for the East Coast of the United States. Dorian joins Michael, Harvey, and Maria as billions dollar+ weather events that took the lives of people. Their names will be retired, and people will ask how they are connected to climate change. These storms certainly exhibited characteristics of processes reported in the scientific literature: Rapid intensification,[40] stalling,[41] and higher sea levels within storm surges.[42] However, the process of extreme weather attribution must be carefully considered without hype, speculation, and social media debates.

Many of these same questions are posed about recent heat waves, fires in Alaska, flooding, and drought activity impacting agricultural lands. In 2016, I served as a co-author on a study conducted by The National

[39] https://www.forbes.com/sites/marshallshepherd/2019/09/03/why-hurricane-dorian-has-stalled-over-the-bahamas/#1447fcfa224a.
[40] https://www.gfdl.noaa.gov/global-warming-and-hurricanes/.
[41] https://www.nature.com/articles/s41612-019-0074-8.
[42] https://toolkit.climate.gov/topics/coastal/storm-surge.

Academies of Sciences, Engineering, and Medicine[43] called "Attribution of Extreme Weather Events in the Context of Climate Change." A key finding from the report is that scientific studies are able to provide some degree of attribution with moderate to high confidence for certain weather events (Figure 1). I should caution that it is irresponsible to link singular events to climate change with the "ill-posed" question, "Was that hurricane, flood, or drought caused by climate change?" However, confidence is growing that the fingerprint of climate change is firmly seen in extreme temperature events (e.g., extreme heat and lack of extreme cold events). Attribution of extreme rainfall, hydrological drought, and hurricanes are found in the moderate confidence category. There is little to no confidence in attribution of tornadic storms or Nor'easter-Mid-latitude cyclone type storms.

The American Meteorological Society recently weighed in on Attribution in its Information Statement on Climate Change:[44]

> "Heavy precipitation (e.g., maximum daily precipitation in consecutive 5-year segments) has increased in both intensity and frequency since 1900, especially in the eastern half of the United States and notably in the Northeast. Areas that receive limited precipitation, sometimes called drylands, are increasing in area. The combination of warmer temperature and reduced precipitation in some regions has increased the risk of drought and drought-related impacts. There is evidence that wildfire seasons are increasing globally and areas where wildfires occur are expanding...The number and intensity of Atlantic hurricanes have both increased since the early 1980s, but much of this increase may be due to natural variability of the atmosphere and ocean. Furthermore, there is little trend or even a decrease in hurricane activity in other ocean basins, so the global trend, if there is one, is not clear. There is evidence that ocean warming is providing more energy to make hurricanes more intense....There is no sign of an increase in the most violent U.S. tornadoes (those rated EF4 or EF5 on the Enhanced Fujita

---

43 https://www.nap.edu/catalog/21852/attribution-of-extreme-weather-events-in-the-context-of-climate-change.

44 https://www.ametsoc.org/index.cfm/ams/about-ams/ams-statements/statements-of-the-ams-in-force/climate-change1/.

Scale). However, there is evidence that annual U.S. tornado activity has become more variable since the 1970s, with larger tornado outbreaks separated by longer periods of below-average tornado frequency."

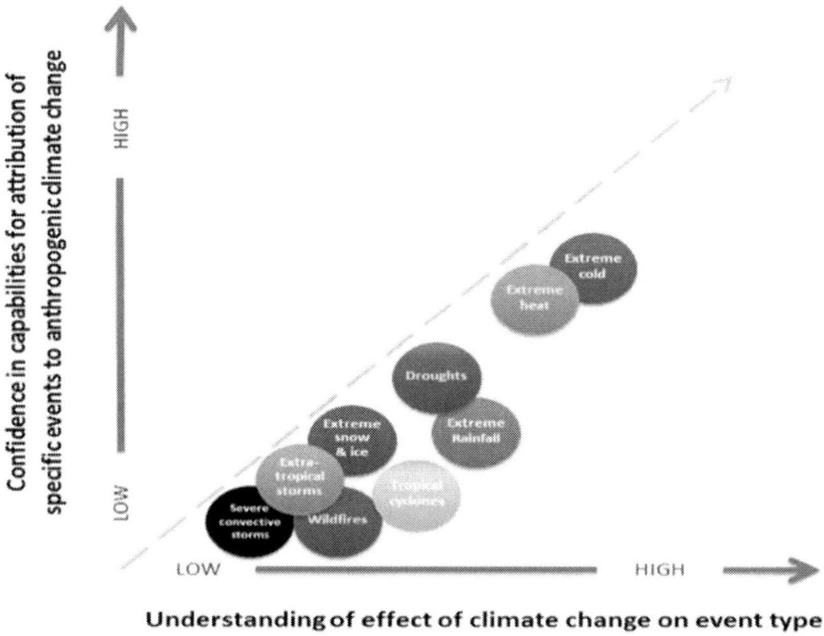

Figure 1. Confidence, based on current literature, in extreme weather linkages to climate change (NAS, 2016).

Our methodology for confidence in the National Academies study relied on sound physical principles, consistent observational evidence, and the ability for numerical models to reproduce the event. These "three legs of the stool" served as the benchmarks in our review of studies and scientific methodologies. Many events ranked lower on the confidence scale may ultimately end up higher as the science evolves. There is emerging literature that may increase our confidence going forward. For example, the best consensus on hurricanes and climate change points to less frequent but stronger storms. However, emerging studies find that hurricanes are stalling more frequently as we witnessed with Imelda (2019), Harvey (2017), Dorian (2019), and Florence (2018).

Figure 2. Rainfall footprint of Hurricane Dorian (2019) as measured by NASA satellite-based instrumentation associated with the Global Precipitation Measurement Mission (GPM).

Indeed, climate changes naturally and always has. I assure you that every climate scientist of significance understands this fact. It is often amusing when people remind degreed scientists of this fact in the mall, at dinner or on social media. It is not an "either/or" proposition. It is an "and" proposition. Grass grows naturally, but it grows very differently when soil is fertilized. Our climate is being fertilized by increasing levels of greenhouse gases (GHGs) in our atmosphere, and I want to be crystal clear here that GHGs are the dominant source for warming.[45]

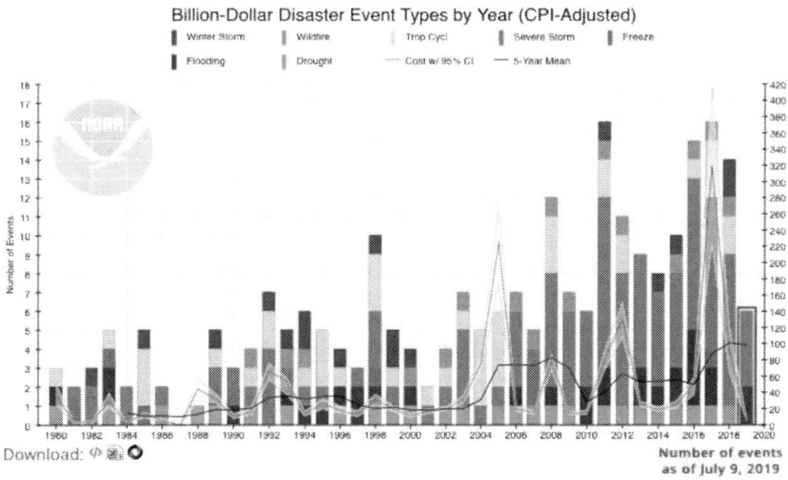

Figure 3. Trends in billion-dollar disasters (Source: NOAA-NCEI).

Suspended particulates (aerosols)[46] and land use changes[47] related to deforestation, urbanization, and agriculture also play important roles in changes being observed from local to global scale. All aspects of the Earth's system, including weather, are responding. For example, NOAA recently updated what constitutes a 100 or 1000 year event in Texas because of changes in rainfall intensity.[48] This will have implications on the National Flood Insurance Program and infrastructure design.

---

[45] http://www.climatechange2013.org/images/figures/WGI_AR5_Fig8-17.jpg.
[46] https://www.sciencedirect.com/science/article/pii/B9780444635242000270.
[47] https://science.sciencemag.org/content/310/5754/1625.summary.
[48] https://www.noaa.gov/media-release/noaa-updates-texas-rainfall-frequency-values.

Messaging Emergencies in the Extreme Weather-Climate Change Connections

In 2018, there were 39 billion-dollar+ disasters. According to the annual report of Aon, a leading insurance broker, insured dollars totaled $90 billion, which is the fourth highest inflation-adjusted number of such events on record.[49] Of those events, the United States tallied 16 of them, the most of any country and second-highest on record according to Dr. Jeff Masters.[50] The U.S. experienced 20 billion-dollar+ weather disasters in 2017. As a reference, there were only 28 billion dollar events within the entire decade of the 1980s according to data from the National Oceanic and Atmospheric Administration and NBC News.[51] Trends certainly indicate an uptick in both weather disasters (Figure 3) and infrastructure.[52] This suggests that the United States faces amplified and compounded risks from extreme weather, population pressures, and wealth.

At a time in which national emergencies are being discussed in Washington, D.C., I argue that climate change has reached the level of a national emergency or crisis. It is imperative that we speak on the topic in such terms rather than distant or abstract references that may not resonate with the average U.S. citizen. I like polar bears and butterflies, but climate messaging must firmly be anchored in how extreme events affect our water supply, public health, infrastructure, energy systems, food supply, and national security.

In 2018, Hurricane Michael devastated my home state of Georgia, and my heart ached for the hard-working families whose pecans, peanuts, bell peppers, and cotton were lost.[53] As tragic as the losses were for my state, it is important that we connect the dots for people that when they in their

---

[49] http://thoughtleadership.aonbenfield.com/Documents/20190122-ab-if-annual-weather- climate-report-2018.pdf.

[50] https://weather.com/news/weather/news/2019-01-23-39-billion-dollar-disasters-2018-first-2019.

[51] https://www.nbcnews.com/politics/meet-the-press/extreme-weather-events-costs-are-piling-n1054576.

[52] https://journals.ametsoc.org/doi/full/10.1175/BAMS-D-17-0184.1.

[53] http://agr.georgia.gov/GDA-Hurricane-Response/media/2018-Hurricane-Michael-Georgia-ag-impacts.pdf.

personal budgets. The 2018 National Climate Assessment report warned that climate change could have a significant negative impact on the U.S. economy and its Gross Domestic Product.[54]

My approach to communicating the climate crisis to the public, stakeholders, and policymakers is to utilize analogies, familiar stories, and common language. I recently served on a panel with former Georgia Congressman Lindsay Thomas. During his remarks, the former Congressman cited a scripture from the same text that I read:

> All streams flow into the sea, yet the sea is never full. To the place the streams come from, there they return again. – Ecclesiastes 1:7 (New International Version)

This text perfectly captures the hydrological or water cycle that most of us learned about in 4th grade. It is a cycle in perfect harmony and balance until human intervention started to disrupt it. More intensive downpours associated with climate-enhanced moisture availability, sustained downpours and melting snowpacks are overwhelming flood protection and stormwater management systems. The Environmental Protection Agency website discussion[55] from 2017 warned:

> "In many areas, climate change is likely to increase water demand while shrinking water supplies. This shifting balance would challenge water managers to simultaneously meet the needs of growing communities, sensitive ecosystems, farmers, ranchers, energy producers, and manufacturers. In some areas, water shortages will be less of a problem than increases in runoff, flooding, or sea level rise. These effects can reduce the quality of water and can damage the infrastructure that we use to transport and deliver water."

---

[54] https://nca2018.globalchange.gov.

[55] https://19january2017snapshot.epa.gov/climate-impacts/climate-impacts-water-resources_ html.

Water availability and infrastructure resiliency are issues that would seem immune from partisan ideology. Water is essential to life and doesn't understand the concept of "liberal" or "conservative."

I am a climate scientist, but I am a husband, father, and community citizen like other Americans. We care about our children, their health, and our great nation. The graphs, charts, and jargon can be a distraction. The climate emergency affecting families is real and must be conveyed in relatable terms.

Some Ideas to Move Forward

I am a scientist trained to understand how our atmosphere works and how to predict its changes. My expertise does not lie in the 2, 4 or 6 year political cycles that you have to think about. However, I am grateful for this opportunity to serve my country through my expertise. I would like to offer the following ideas as steps forward:

Keep Scientific Observation and Modeling Capacity Robust

Diverse observations are essential when meteorologists are diagnosing the atmospheric and oceanic conditions that may lead to rapid intensification of a hurricane or sustained flooding. NOAA, NASA, and other satellite datasets are critical for diagnosing extreme weather in real-time, but these datasets are also assimilated into our numerical weather prediction models. Denial studies have shown, for example, that the forecast for Hurricane Sandy would have been terrible if satellite datasets were left out of the models.[56] I chair NASA's Earth Science Advisory Committee. NASA along with federal, international, and private partners continue to provide vital Earth system observations of Earth. NASA is currently implementing the Decadal Survey plan recommended by the National Academies. The agency has initiated several studies and issued

---

[56] https://www.noaa.gov/media-release/noaa-updates-texas-rainfall-frequency-values.

calls to the community in support of ensuring that vital observations of the weather-climate system continue and are advanced.[57]

NOAA and the National Center for Atmospheric Research (NCAR) recently signed agreements to establish, according to a federally issued press release,[58] "*a new partnership to design a common modeling infrastructure (EPIC) that will be transparent and easy to access and use by public and private researchers, including academia and industry. By leveraging efficiencies and synergies, reducing duplication of effort, and creating shared model repositories, future research advances will more quickly benefit the public through better weather and climate forecasts.*" Challenges with rainfall forecasts in Tropical Depression Imelda in Texas (2019) and rapid intensity changes observed with Hurricane Michael (2018) affirm the need for the EPIC framework. I believe EPIC is a positive step to ensure a more nimble and responsive U.S. weather model capability as we keep pace with or attempt to surpass other global modeling efforts.

To improve extreme weather prediction in the 5 to 14 day time window, we must also aggressively maintain our high performance computing and data assimilation resources. Such capacity will also improve emerging sub-seasonal to seasonal forecasting capabilities critical to agricultural, energy, and other applied needs. We need the fastest supercomputers available to accommodate this generation's volume of observational data and computational codes needed to advance our predictive capabilities. We know what is needed. We just need the continued support.

Advances in small or cube satellite technologies and precision instrumentation aboard research aircraft at NASA and NOAA enable nimble and efficient measurements required to address specific aspects of extreme weather prediction. For example, hurricane track forecasts have steadily improved in recent decades while intensity forecasts have lagged

---

[57] https://science.nasa.gov/earth-science/decadal-surveys.

[58] https://www.noaa.gov/media-release/noaa-and-ncar-partner-on-new-state-of-art-us-modeling-framework.

behind.[59] Scientific interrogation of the processes within the eyewall of hurricanes and beneath the surface will likely move us forward.

It is also important that we find a compromise on the use of the electromagnetic spectrum used for telecommunications and weather so that vital observations assimilated into weather prediction models are not degraded.[60] The efficacy of modern day weather models is highly dependent upon temperature, water vapor, and other satellite-derived fields.

As important as the technology of extreme weather forecasting is to the enterprise, it is clear that a "good forecast" easily becomes a "bad forecast" if people are not consuming, understanding or acting upon the information. Investments in social sciences (psychology, communication, sociology, equity studies, economics) at the intersection of extreme weather will hopefully help prevent tragedies that could be avoided. As we move toward a Weather Ready Nation, science increasingly plays a role as a decision support service with regard to how use-inspired research initiates, O2R, and how to frame forecasts for utility services, state and municipal governments, emergency management, water management, public health, and so forth.

Learn from Best Practices in Regional or Stakeholder Efforts

In Georgia, through the support of the Ray C. Anderson Foundation, I am participating in a unique consortium called the Georgia Climate Project.[61] The effort was called out by the 2018 National Climate Assessment as a potential best practice to be replicated in other states and jurisdictions. Our goals are: (a) Synthesizing what is known and analyzing what is not in order to improve understanding of climate impacts and solutions in Georgia, (b) Fostering a constructive, nonpartisan discussion about how climate change affects Georgia and what can be done about it, (c) Working with partners to enable Georgians to take practical steps to

---

[59] https://www.forbes.com/sites/marshallshepherd/2015/08/28/why-is-tracking-a-hurricane-easier-than-predicting-its-intensity/#3c496e691721.

[60] https://www.wyden.senate.gov/news/press-releases/wyden-and-cantwell-to-fcc-dont-ignore-nasa-noaa-and-navy-concerns-on-5g-auction.

[61] https://www.georgiaclimateproject.org.

respond to climate change and its impacts, and (d) Bringing together experts working to understand and act on climate. We will host the Georgia Climate Conference in November to assess where these efforts are and to move forward. While such regional actions can move the needle on science and policy, they do not replace the vital federal role in providing observational platforms, models, research grants, policy and international diplomacy.

Understand Risk, Vulnerability, and Resiliency

A common question that I often receive is "So what do we do about climate change and extreme weather?" Mitigation strategies centered on reducing carbon emissions have been the dominant themes along with adaptation strategies. My focus is increasingly on aspects of risk, vulnerability, and resiliency. I am particularly concerned about the "Extreme Weather-Climate Gap." Disparities in income, social status, and other factors that lead to marginalized groups mean that hurricanes, floods, or heat waves have disproportionately adverse impacts on certain populations. Such vulnerable populations have elevated risks in terms of health, resiliency, and economic well-being when facing extreme weather events.[62]

This is clearly apparent when you look at a cross section of how communities recovered after Hurricane Katrina and Harvey, respectively. The other aspects of resiliency involve urban and rural infrastructure. AT&T, for example has been a leader in its efforts to address climate resiliency, in part because they understand their own vulnerabilities.[63] The U.S. Navy is constantly thinking about resiliency as its facilities struggle with sea level rise and a generation of more intense hurricanes.[64] More locally, major cities and small, rural towns face degradation to transportation, water-delivery, and agricultural systems. The Institute for Resilient Infrastructure Systems (IRIS) at the University of Georgia, of

---

[62] https://www.sciencedirect.com/science/article/pii/S0143622815000909?via%3Dihub.
[63] https://about.att.com/newsroom/2019/climate_resiliency_community_challenge.html.
[64] https://www.gao.gov/assets/700/699679.pdf.

which I am a part of, is starting to think about such issues but will require adequate resources to think about such challenges in innovative ways.[65]

Concluding Thoughts

In summary, anthropogenic climate change superimposed onto the naturally varying climate system has created an era of extreme weather events foreseen by past studies in the scientific literature. Climate scientists, in most cases, typically try to be measured with messaging and objective in their analyses. However, the climate system itself is sounding the alarms with the current generation of heat waves, floods, and storms.

- Attribution studies affirm that the influence of climate change is found in contemporary extreme weather events affecting the United States. They will continue to advance to better understand the connections between climate change and extreme weather events with the goal of improving predictive capability and building in resiliency to our societal frameworks.
- Extreme weather events impact people, infrastructure and processes more than "average" events. Our agricultural, energy, water, transportation, public health, and national security institutions are at risk, and this ultimately affects American households and beyond. We must message the risks and impacts with such context in mind rather than distant and abstract concepts.
- Robust science and technology resources must be sustained at local to federal levels.
- We must keep a close eye on the well-being of vulnerable populations and disparities in climate resiliency.

---

[65] http://iris.uga.edu.

Biographical Sketch Dr. J. Marshall Shepherd University of Georgia

Dr. J. Marshall Shepherd is a leading international expert in weather and climate and is the Georgia Athletic Association Distinguished Professor of Geography and Atmospheric Sciences at the University of Georgia. Dr. Shepherd was the 2013 President of American Meteorological Society (AMS), the nation's largest and oldest professional/science society in the atmospheric and related sciences. Dr. Shepherd serves as Director of the University of Georgia's (UGA) Atmospheric Sciences Program and Full Professor in the Department of Geography where he was a previous Associate Department Head. Dr. Shepherd is also the host of The Weather Channel's Award-Winning show Weather Geeks, a pioneering Sunday talk podcast/show and a contributor to Forbes Magazine. Dr. Shepherd is the 2019 Recipient of the AGU Climate Communication Prize and the 2018 recipient of the prestigious AMS Helmut Landsberg Award for pioneering and significant work in urban climate. In 2017, he was honored with the AMS Brooks Award, a high honor within the field of meteorology. Ted Turner and his Captain Planet Foundation honored Dr. Shepherd in 2014 with its Protector of the Earth Award. Prior recipients include Erin Brockovich and former EPA Administrator Lisa Jackson. He is also the 2015 Recipient of the Association of American Geographers (AAG) Media Achievement award, the Florida State University Grads Made Good Award and the UGA Franklin College of Arts and Sciences Sandy Beaver Award for Excellence in Teaching. In 2015, Dr. Shepherd was invited to moderate the White House Champions for Change event. He is an alumni of the prestigious SEC Academic Leadership Fellows program. Prior to UGA, Dr. Shepherd spent 12 years as a Research Meteorologist at NASA-Goddard Space Flight Center and was Deputy Project Scientist for the Global Precipitation Measurement (GPM) mission, a multi-national space mission that launched in 2014. President Bush honored him on May 4th 2004 at the White House with the Presidential Early Career Award for pioneering scientific research in weather and climate science. Dr. Shepherd is a Fellow of the American Meteorological Society. Two national magazines, the AMS, and Florida State University have also recognized Dr. Shepherd for his significant contributions. Dr. Shepherd was the 2016 Spring Undergraduate Commencement speaker at his 3-time Alma Mater, Florida State University. He was also the 2017 Graduate Commencement speaker at the University of Georgia.

Dr. Shepherd is frequently sought as an expert on weather, climate, and remote sensing. He routinely appears on CBS Face The Nation, NOVA, The Today Show, CNN, Fox News, The Weather Channel and several others. His TedX Atlanta Talk on "Slaying Climate Zombies" is one of the most viewed climate lectures on YouTube. Dr. Shepherd is also frequently asked to advise key leaders at NASA, the White House, Congress, Department of Defense, and officials from foreign countries. In February 2013, Dr. Shepherd briefed the U.S. Senate on climate change and extreme weather. He has also written several editorials for CNN, Washington Post, Atlanta Journal Constitution, and numerous other outlets and has been featured in Time Magazine, Popular Mechanic s, and NPR Science Friday. He has over 90 peer-reviewed scholarly publications. Dr. Shepherd has attracted $3 million dollars in extramural research support from NASA, National Science Foundation, Department of Energy, Defense Threat Reduction Agency, and U.S. Forest Service. Dr. Shepherd was also instrumental in leading the effort for UGA to become the 78th member of the University Corporation for Atmospheric Research (UCAR), a significant milestone for UGA and establishing UGA's Major in Atmospheric Sciences.

Dr. Shepherd currently chairs the NASA Earth Sciences Advisory Committee and was a past member of its Earth Science Subcommittee of the NASA Advisory Council. He was a member of the Board of Trustees for the Nature Conservancy (Georgia Chapter), National Oceanic and Atmospheric Administration (NOAA) Science Advisory Board, Atlanta Mayor Kasim Reed's Hazard Preparedness Advisory Group United Nations World Meteorological Organization steering committee on aerosols and precipitation, 2007 Inter-governmental Panel on Climate Change (IPCC) AR4 contributing author team, National Academies of

Sciences (NAS) Panels on climate and national security, extreme weather attribution, and urban meteorology. Dr. Shepherd is a past editor for both the Journal of Applied Meteorology and Climatology and Geography Compass, respectively.

Dr. Shepherd received his B.S., M.S. and PhD in physical meteorology from Florida State University. He was the first African American to receive a PhD from the Florida State University Department of Meteorology, one of the nation's oldest and respected. He is also the 2nd African American to preside over the American Meteorological Society. He is a member of thc AMS, American Geophysical Union, Association of American Geographers (AAG), Sigma Xi Research Honorary, Chi Epsilon Pi Meteorology Honorary, and Omicron Delta Kappa National Honorary. He is also a member of the Alpha Phi Alpha Fraternity, Inc. and serves on various National Boards associated with his alma mater. Dr. Shepherd co-authored a children's book on weather and weather instruments called D r. Fre d's Weather Watch. He is also the co-founder of the Alcova Elementary Weather Science Chat series that exposes K-5 students to world-class scientists. Dr. Shepherd is originally from Canton, Georgia. He is married to Ayana Shepherd and has two kids, Anderson and Arissa.

Chairwoman JOHNSON. Thank you. Dr. Done.

TESTIMONY OF DR. JAMES DONE, PROJECT SCIENTIST III AND WILLIS RESEARCH FELLOW, CAPACITY CENTER FOR CLIMATE AND WEATHER EXTREMES, MESOSCALE & MICROSCALE METEOROLOGY LAB, NATIONAL CENTER FOR ATMOSPHERIC RESEARCH

Dr. DONE. Good morning, Chairwoman Johnson and Ranking Member Lucas. Thank you for this opportunity to testify today at this important hearing.

So the United States is no stranger to extreme weather events, but the impacts from recent events have been unprecedented. And in fact we're in a new era of extreme weather. So we've experienced deadly heat. We've seen the devastating floods from recent Hurricanes Harvey, Florence, and just last week Tropical Storm Imelda, as mentioned earlier. And we've experienced the tragedy of fastmoving intense wildfire.

So what's causing these changes? Well, the impacts from extreme events are due to characteristics of the weather events themselves but also due to characteristics of what's in harm's way.

So now into climate change, a growing and pervasive risk multiplier. So, sure, our rising populations have contributed strongly to our rising impacts, but as a physical scientist I can see that the events that cost money such as flooding rains have increased.

So as we saw the images of Hurricane Dorian recently, that just showed the potential for catastrophic intersection between a record-breaking weather event and our rising population, shown there in the night lights. So given that today's atmosphere is warmer and more moist than it used to be, it's inconceivable that today's weather has not changed. Indeed, our droughts are hotter, wildfires are larger, and our heavy rain events are even heavier.

So what of the future? So 1 or 2 degrees temperature rise sounds fairly small, but the impacts are expected to be anything but. So we expect the rains to become even heavier. Most hurricane scientists will tell you they expect faster winds, heavier flooding rains, and more extensive storm surge inundation.

Now consider wildfire. So already California is already one of the most flammable regions on Earth. The aspect of climate change we understand most on wildfire is the impact of our warming atmosphere. It demands more moisture out of the vegetation. This desiccated vegetation leads to more intensely burning wildfires that are fundamentally of a different character to the ones we see today.

Now, the U.S. has world-class science and technology, so this includes sponsorship of the National Center of Atmospheric Research by the National Science Foundation, but we lack key understanding of the most damaging events. And perhaps more importantly there's a disconnect between our advancing science and societal benefit. And in this area of changing extremes, it's more important than ever to have solid, well-communicated, short-term forecasts and robust risk management strategies. So I believe there's huge gains to be had by a deep integration of our advancing science with risk management.

So allow me to just give you an example from my recent experience as a Willis Research Fellow. I collaborate with the reinsurance industry. So through our interactions I learned that hazard risk commonly assumes that these extreme events don't know about each other, but I've seen in the data that some events, they're like buses. You wait for ages and then three come along at once. So scientists are excited to know how this can advance forecasting, and risk managers are interested in designing away this vulnerability.

So my second example comes from building codes. Some work with I did with economists showed that for every dollar you spend building to code, you can expect $2 to $8 back in reduced losses. So this is clearly sound economic policy. And it was demonstrated to dramatic effect when Michael last year we saw homes that were not built to code were completely destroyed. Homes that were built to code suffered relatively minor damage.

So to pursue this deep integration of science and risk management, the Federal Government has a vital role to play. So the new NOAA, NCAR, and community weather and climate modeling partnership, together with the EPIC, really serves as a model for how this should happen. It directs science squarely in the needs of society.

And in terms of bolstering the science, as was mentioned earlier, we need to sustain our resources for continued observational platforms such as the Oklahoma mesonet, sustained computational infrastructure, the creation of a national data set of extreme events, and also sustained research grants to analyze and develop understanding.

So, in conclusion then, let me reiterate the importance in this new era of extremes for solid, well-communicated, short-term forecasts and robust risk management. So thank you again, Chairwoman Johnson and Ranking Member Lucas, for this opportunity.

[The prepared statement of Dr. Done follows:]

WRITTEN TESTIMONY OF DR. JAMES M. DONE, DEPUTY DIRECTOR, CAPACITY CENTER FOR CLIMATE AND WEATHER EXTREMES, NATIONAL CENTER FOR ATMOSPHERIC RESEARCH, AND WILLIS RESEARCH FELLOW, BEFORE THE HOUSE COMMITTEE ON SCIENCE, SPACE, & TECHNOLOGY, UNDERSTANDING, FORECASTING, AND COMMUNICATING EXTREME WEATHER IN A CHANGING CLIMATE, SEPTEMBER 26, 2019

Good morning Chairwoman Johnson and Ranking Member Lucas. I am Dr. James Done, deputy director of the Capacity Center for Climate and Weather Extremes at the National Center for Atmospheric Research. Thank you for this opportunity to participate in this important hearing. I'm a physical scientist specializing in extreme weather. In my role of Senior Academic Fellow of the Willis Research Network I collaborate with the reinsurance industry to advance our understanding of extreme weather and effective risk management. It's my pleasure to testify today on the state of the science of extreme weather under climate change and to outline opportunities to bolster our science, risk management, and protection of lives and property.

Executive Summary

Extreme weather is the main vehicle that delivers the impacts of climate change. We have already detected a signal of climate change in deadly heatwaves and flooding rains. And further changes in extreme weather are anticipated. Whereas the incremental warming of our earth system may seem small, on-the-ground impacts are anything but. In addition to local devastation, extreme weather impacts reverberate through our natural, physical, and social systems.

The U.S. has excelled at producing breakthrough research advances and technologies that benefit science, human safety, economic prosperity, and national security. It's now more important than ever to have effective communication of solid short-term forecasts, and robust longer-term risk management strategies. A deeper integration of science and practice is urgently needed to strengthen protection of lives and property, ensure military readiness, and sustain economic competitiveness. Our choice is clear: Implement solutions, or face greater catastrophes.

Introduction

Thank you for the opportunity to discuss this urgent topic. I shall begin outlining the state of today's extreme weather and how it's different from the past. Then I'll outline what we can expect in the future. I'll close with some recommendations for how we can respond by bolstering our science and risk management practice.

We have entered a new era of extreme weather. While the U.S. is no stranger to extreme events, impacts from recent events have been unprecedented. We've experienced the devastating floods brought by Harvey, Florence and Imelda. We've baked under deadly heat. We've endured multi-year droughts. And we've experienced the tragedy of fast-moving, intensely hot wildfires. New forms of extreme weather have appeared. The Carr Fire near Redding, California in 2018 produced a long-lived tornado with winds up to 150mph, uprooting trees and downing power lines. Alaskan coastlines previously protected by sea-ice are now seeing erosion from Arctic cyclones. What is causing these rising impacts?

Extreme weather has changed. So have we. Impacts arise from the intersection of extreme weather with our rising populations and increasingly interconnected systems. Enter climate change, a pervasive and growing risk multiplier. All these factors contribute to impacts. Questions of whether climate change caused an event miss this point. Sure, our shifting demographics are a strong contributor. And on top of that, evidence is building for a climate change signal in our rising losses. As a physical scientist I know that the events that cost money and lives, such as coastal flooding from sea level rise or flooding rains from intense thunderstorms, have increased.

Figure 4. A striking visual of Category 5 Hurricane Dorian on Sept 2, 2019 showing the exposure of our rising population to todays' record-breaking extreme weather.

Our challenge then, is to understand these factors, and use that knowledge to inform actions to reduce future losses and maximize potential benefits. It's now more crucial than ever to fully integrate our science with our risk management practice to strengthen protection of lives and property, and support military readiness and economic competitiveness.

We Live in a New Era of Changing Extreme Weather

Today's warmer atmosphere drives more moisture out of our vegetation, soil, and oceans. We have a fundamental scientific principle that tells us that the moisture in the atmosphere increases by about 6-7% for every degree Celsius of warming. What does this mean for extreme weather? Given that some warming has already occurred it's inconceivable that today's weather, that operates within this warmer, moisture-rich environment, has not changed. Indeed, the hand of climate change has already been detected across a range of extreme weather phenomena.

Heavy rainfall for example has followed the same rate of increase as moisture. We have detected increases in heavy precipitation across the U.S. with the largest increases in the Northeast. There is also some evidence that rain rates may be rising even faster than the rate of moisture increase. Recent studies found the climate change contribution to Hurricane Harvey's rainfall could have been as much as 20%. But more work is needed to understand this rather alarming possibility.

What does this mean for our aging storm water infrastructure? Figure 2 compares two photos of the same street corner in downtown Pittsburgh, PA taken almost 100 years apart. Much has changed in 100 years. What hasn't changed is the storm water infrastructure. Can our aging storm water infrastructure adequately collect and convey today's deluges?

Downtown Pittsburgh, 1912

Downtown Pittsburgh, 2010

Figure 5. Two photos of the same street corner in downtown Pittsburgh, taken almost 100 years apart. (PG Archive, left), (Pittsburgh Post-Gazatte, right). With thanks to T. Lopez-Cantu (CMU).

Hurricanes are multi-hazard phenomena that combine flooding rains with dangerous surge and destructive winds. What do we know about changes to these other characteristics? Sea levels are higher today than a few decades ago. Today's storm surges ride on top of these higher seas. We don't know yet whether climate change has contributed to changes in North Atlantic hurricane numbers or peak wind speeds. This does not mean that climate change is not playing a role - we just don't have the data to be sure. But most scientists agree that climate change has contributed to a global increase in the fraction of Category 4, 5 hurricanes. Other observed changes include a global slowdown in tropical cyclones, including a trend in stalling storms over our vulnerable coastal populations. But this has not yet been linked to climate change.

The good news is that our ability to forecast hurricanes has improved markedly. Hurricane Dorian is a case in point. It's remarkable to me that we had a Category 5 hurricane 150 miles off the coast of Miami (Figure 1) yet no evacuations in the Miami area were required and people could go about their business as usual. A robust observing network, and vast strides in prediction technologies, both funded by the federal government, paid dividends.

What of our other extreme weather phenomena? The regions favored by tornadoes have shifted east across the central U.S. But we don't yet understand the cause. There is also evidence that tornadoes are clustering in greater numbers, with longer tornado-free periods in-between. But again we don't understand the cause.

Now consider wildfire. California is one of the most flammable places on Earth. While the numbers of wildfires are holding steady, today's fires are much larger than in the past. They are also moving faster and burning more intensely.

The most well-understood climate change impact on wildfire behavior is through warmer summers driving more moisture out of vegetation. This crisp vegetation provides the fuel for when fire weather conditions arise. There is also evidence that Fall rains are starting later. These first rains of the Fall bring critical moisture that shut down the fire season. Take the case of the tragic Camp fire in Paradise, California last November. This fire would almost certainly not have occurred had the Fall rains arrived as usual. Last year's near-record warm summer desiccated the vegetation. While strong winds were a key factor, strong winds in damp forest just don't drive large, fast-spreading wild fire.

What Can We Expect in the Future?

Many people question whether we have reached a new normal of extreme weather?

The answer is emphatically no. There's no evidence to suggest we've reached some kind of stable state of extreme weather. We understand that extreme weather will continue to worsen. And it will combine with demographic and ecological shifts to bring new risks.

Looking to the future, there is mounting evidence that, rather ominously, the more extreme the event, the more intense and more frequent it will become. Take heavy rainfall for example. A recent study by my colleagues found a 4-fold increase in the frequency of the heaviest summer thunderstorm rainfall by the end of this century. Not only does the rainfall rate increase by a 6-7% rise per degree warming, but there is some evidence that the storms may cover greater areas. These compounding effects would double total water volumes falling from these storms, with catastrophic simultaneous multi-watershed flooding. For the case of hurricanes, recent work suggests will we will see more storms reaching the most intense hurricane categories. But more work is needed to ensure this result is robust.

For hurricanes in general most scientists will say they expect faster wind speeds and worse hurricane-related flooding due to heavier rains and stronger storm surges. Unfortunately for us property damage rises about 3 times as fast as the rise in wind speed. So even modest increases will lead to large changes in damage.

Changes to other impactful characteristics of hurricanes are less certain. We don't know, for example, whether we'll see more stalled hurricanes on the U.S. coast. Nor do we know how the likelihood of very large hurricanes will change. Likewise, we don't understand how hurricane-spawned tornadoes may change.

Looking at other extreme weather, evidence is mounting for future changes in large hail. The warmer and more-moist atmosphere provides more fuel for hail-generating thunderstorms. There is some evidence for an overall increase in hail size. But the U.S. Gulf States may see fewer large hail events due to the suppressing effect of elevated freezing heights. For tornadoes, it's not clear whether climate change will make them stronger or more frequent. This is not because there will be no change, but because we don't have the understanding or data to be sure.

Fire seasons are getting warmer, meaning that tinder dry vegetation will become more common, all other things being equal. But what about precipitation? Will dry Falls become commonplace, thereby extending fire seasons? There is evidence emerging that Western U.S. winter precipitation may concentrate into the heart of winter, at the expense of Spring and Fall.

There is a clear need to deepen our understanding of how these key damaging characteristics of weather extremes may change. We need to stay ahead of all possible event scenarios. Hurricanes exploding into life right along the U.S. coast is just one example of possible future hurricane behavior that would seriously compromise our ability to issue warnings and enact emergency responses.

Innovating Our Science and Risk Management

Given this era of unprecedented high-impact weather events, and the expectation of worse to come, it's more important than ever to have solid short-term forecasts and robust longer-term risk management strategies. This is essential to protect lives and property, ensure viable populations in our coastal regions, and to ensure military readiness and economic competitiveness.

The weather enterprise, including sponsorship of the National Center for Atmospheric Research by the National Science Foundation, has transformed our knowledge, research infrastructure, and community capacity for extreme events. Despite these leaps in capacity, we still suffer from poor understanding of some of the most high-impact weather events, and a disconnect between scientific advances and societal benefit.

Our changing extreme weather is fundamentally a question of risk assessment. We need to understand what impacts are possible, and how likely they are. This requires us to focus on the impacts to understand the science that needs to be done. Impacts are driven by a confluence of factors: our demographics, characteristics of the weather events, with climate change an all-pervasive and growing threat multiplier. Our response, then, must consider all these ingredients of risk and their interactions at the same time. This requires deep integration of science with risk management.

My experience as a Willis Research Fellow taught me that there are huge potential gains in our ability to protect life and property through deep integration of science and risk management. Allow me to give a current example.

While discussing risk management with my colleagues at Willis Towers Watson I learned that hazard risk is commonly based on the assumption that extremes are independent of each other. But I've seen in the data that extreme events know about each. Some are like buses – you wait forever then three arrive at once. These groups of extremes bring potential vulnerabilities for risk management. For example, water managers up and down the U.S. West Coast are all too familiar with parades of winter storms that raise flood risk and stress water management practice. Scientists are excited about the potential for this understanding of connections between extreme events to lead to better forecasts. In turn, risk managers are looking for ways to innovate practice to design away this vulnerability.

My recent work exploring the effectiveness of building codes provides another example of the value of integration. The devastation caused by Hurricane Andrew in 1992 was largely attributed to poor construction. In response the state of Florida retrofitted their building code to mitigate future hurricane losses. But how effective is this code? I conducted a study with economists to explore that question. We found up to a 72% reduction in losses for homes built to the new code, thereby demonstrating that the code is indeed very effective.

But it costs money to build to code so is it worth the expense? Our research found that for every dollar spent building to code, you get between 2 and 8 dollars back in mitigated losses. My colleagues found similar levels of cost benefits of building codes against tornado risk in Oklahoma.

If Hurricane Michael's impacts on the Florida Panhandle (Figure 3) and into Georgia last year are indicative of the future then the benefits of effective building codes will continue to rise. Yet these benefits will only accrue if we understand the future conditions to which codes should be responsive to. Without codes designed for future conditions, deteriorating levels of protection may lead to people choosing to leave impacted areas. Intensifying drought, wild fires, and hurricanes may all begin to enter people's decisions of where to live.

Figure 6. Mexico Beach, FL after Hurricane Michael (2018). Older homes not built to code are completely destroyed. But newer homes built to code suffered relatively minor damage. (Dronebase/Reuters, Handout/Reuters).

Not only do building codes mitigate losses, but can also make insurance more affordable thereby closing the so-called protection gap. Better understanding of the range of future extreme weather events is critically needed to gauge the effectiveness of current codes, analyze the effects of policy decisions, and inform retrofitting our nation for the future.

These experiences lead me to ask: What other opportunities are we missing to bolster our science and practice? Finding these opportunities won't happen by accident. It requires leadership.

The federal government has a vital role to play in discovering and pursuing these opportunities to better protect lives and property. The new NOAA and NCAR weather and climate community modeling partnership serves as a model for accelerating our science and guiding it in the direction of society's most pressing needs. In addition, the establishment of the Earth Prediction Innovation Center promises to further accelerate the weather enterprise. Furthering these efforts, and funding other opportunities for success are urgently needed. Let me outline one immediate opportunity we can't afford to miss out on.

Scientists are at the cusp of a breakthrough in our ability to simulate weather over many decades at a level of granularity not previously possible. With continued federal support we'll have the capacity to simulate many hundreds of hurricanes for example, or many hundreds of flooding rain events over the Mid-West. All of them physically possible in today's climate. Of course, these simulations will not be without error, but they have the potential to be incredibly useful if developed in concert with risk managers.

Resources are required to build capacity for sustained, coordinated and integrated efforts across science and practice. This includes providing continued support in the form of observational platforms, computational infrastructure, and research grants. One specific need is the establishment of a national multidisciplinary dataset of extreme weather impacts. This vital national asset will catalyze scientific understanding and risk management solutions at the forefront of societal needs. In this era of changing extremes, integrated science and practice is not optional.

Conclusion

Extreme weather impacts continue to increase. A confluence of factors drives these impacts, including characteristics of what's in harm's way and the weather events themselves. Climate change acts as a pervasive and compounding threat multiplier. Indeed, the hand of climate change has already been detected in some extreme weather phenomena. We understand that extreme weather will continue to worsen. And it will combine with demographic and ecological shifts to bring new risks.

This era of changing risk demands solid and well-communicated short-term forecasts and robust longer-term risk management strategies. A deeper integration of science and practice is urgently needed to strengthen protection of lives and property, and ensure military readiness and economic competitiveness. Extreme weather is inevitable, extreme weather catastrophes don't have to be.

Dr. James Done, Project Scientist III and Willis Research Fellow, Capacity Center for Climate & Weather Extremes, Mesoscale & Microscale Meteorology Lab, National Center for Atmospheric Research.

Dr. Done is a Senior Willis Fellow and deputy director of the Capacity Center for Climate and Weather Extremes at the National Center for Atmospheric Research. Dr. Done works with stakeholders from the energy, water and insurance sectors to understand future weather and climate extremes and their impacts. Examples of recent work include; assessing future hurricane impacts on the offshore energy industry, exploring the value of decadal climate prediction for water resource and flood risk management, and understanding the drivers of hurricane wind losses. Dr. Done received his PhD in meteorology from the University of Reading, UK.

Chairwoman JOHNSON. Thank you. Dr. Sobel.

TESTIMONY OF DR. ADAM SOBEL, PROFESSOR, LAMONT-DOHERTY EARTH OBSERVATORY AND SCHOOL OF ENGINEERING AND APPLIED SCIENCES, COLUMBIA UNIVERSITY, AND DIRECTOR AND CHIEF SCIENTIST, INITIATIVE ON EXTREME WEATHER AND CLIMATE, COLUMBIA UNIVERSITY

Dr. SOBEL. Madam Chair, Ranking Member Lucas, and Members of the Committee, thank you for inviting me here today. I'm Adam Sobel, a Professor and Atmospheric Scientist at Columbia University, and I have a longer statement that I'd like to submit for the record.

Extreme weather is changing due to global warming, but we know more about some kinds of events than others. Heatwaves are the best-understood type of extreme weather event. When any heatwave occurs today, it's probable that global warming made it more likely to occur, more intense once it did occur, or both. On the other hand, we know much less about tornadoes. There have been changes, but we can't yet say with confidence that these changes are caused by warming, nor what we expect in the future. Most kinds of extreme weather fall in between these extremes of understanding and ignorance.

I will focus on hurricanes. Hurricane risk is increasing due to climate change. Storm surge-driven coastal flooding is certainly becoming worse due to sea-level rise. Hurricanes in a warmer climate also produce more rain and stronger winds, though there is still debate on the magnitudes of these changes and to what extent they're already evident. We know little, though, about how hurricane frequency or the total number of storms per year changes with warming. Natural variability makes any gradual human-caused trends hard to detect, and models are inconclusive on this question.

Because other aspects of changes in hurricanes only matter when a hurricane actually occurs, this uncertainty about hurricane frequency limits our ability to predict changes in overall hurricane hazard and risk. But it would be a serious mistake to interpret this uncertainty or other similar uncertainties about exchanging extreme weather risk as license to delay action. Uncertainty is not our friend here.

By analogy, imagine the FBI (Federal Bureau of Investigation) has inconclusive but worrying evidence that some bad people somewhere may be planning an attack. These people are having a meeting, and the FBI has managed to plant a microphone in the room, but it's noisy and the bad people are speaking quietly, making it impossible to hear what they're saying. Would we want the FBI to interpret this uncertainty as meaning there's no need to worry, or would we want them to take whatever reasonable measures they can to prevent the attack given whatever information they do have? I think most of us would want to take action.

The same is true with climate and its consequences for extreme weather. Human-induced climate change is happening. We can't wait until all the uncertainties have resolved. By that point, we will have baked in yet much more warming and extreme weather that we could have avoided with earlier action.

I'd like to end with some recommendations for timely research to close key gaps in our knowledge. We certainly need continued investment in forecasts of both weather and climate, including the observations, models, and methodologies that enable them. A greater gap, though, is research that quantifies the risks from extreme weather and their changes as the climate warms in terms of their impacts on human society, including economic losses, human health impacts, food and energy security, and so on. In particular, I advocate development of a new generation of catastrophe models like those used in the insurance industry to assess risks from extreme weather events but that go beyond existing industry standards by explicitly addressing climate change as a component of the changing risk and by being open source and thus subject to rigorous peer review. And I elaborate this in my written testimony.

Thank you for the opportunity to participate in today's hearing, and thank the Committee and your colleagues on both sides of the aisle and both sides of the Capitol for your support for the Nation's research enterprise. And I'd be pleased to answer any questions or provide additional information. Thank you.

[The prepared statement of Dr. Sobel follows:]

Dr. Adam H. Sobel, Professor, Applied Physics & Applied Mathematics and Earth & Environmental Sciences, Columbia University, before the House, Science, Space, and Technology Committee, House of Representatives, Washington, D.C., September 26, 2019

Madam Chair, Ranking Member Lukas, and Members of the Committee, thank you for inviting me to participate in this morning's hearing on extreme weather. I am Adam Sobel, a professor and atmospheric scientist at Columbia University's Lamont-Doherty Earth Observatory and School of Engineering.

Introduction

In this testimony I will cover three topics:

- A brief overview of the relationship of different extreme weather events to climate change;
- In the case of hurricanes in particular, some of the complexities of their relationship to climate, the sources of uncertainty, and the challenges this poses for communicating and acting on our understanding of the risks they pose; and
- Recommendations for future research, with an emphasis on an expanded view of what the insurance industry calls "catastrophe modeling."

My testimony is based on the peer-reviewed literature, including reports from the Intergovernmental Panel on Climate Change, the National Climate Assessments, and others, as well as being informed by my own research and that of my colleagues at Columbia. I was one of the authors of a 2016 National Academy of Sciences Report *Attribution of Extreme Weather Events in the Context of Climate Change,* which also informs my views.

How Extreme Weather Is Affected by Climate Change: An Overview

Extreme weather is changing as a consequence of human-induced climate change. How it is changing, how quickly, and how well we can detect those changes varies across different kinds of extreme weather events. In my view, there are multiple answers to the question "how are extreme weather events changing"?

Let us first establish some basic concepts. Weather is the instantaneous state of the atmosphere and its evolution over short time scales – days, say. Climate, to take the simplest definition, is the average of the weather over long periods of time. The climate is strongly influenced by external factors, some of which are predictable and act over long periods of time: the position of the earth in its orbit around the sun, the slow circulation of the oceans, and the concentrations of greenhouse gases in the atmosphere. While these factors may control the climate, the internal chaotic dynamics of the atmosphere still give the weather much freedom to fluctuate about that climate. So every weather event, including an extreme one, has many proximate causes, and most of those causes are natural. (Chaos theory teaches us, in fact, that we cannot trace these paths of causality very far back in time in the atmosphere.) So it is never accurate to say that climate change "caused" a single weather event without further qualification. But a change in climate can still change the *probability* that a given type of weather event will occur, or the severity of events in a given class when they do occur. (Event attribution studies, such as described in the 2016 National Academy report, assess those changes in probability or severity for individual events, and investigate to what extent – always much less than 100% - climate change can be held responsible for a given event.) We need to understand those causal links in order to know how to best respond to the reality of human-induced climate change.

Our understanding rests on three distinct sources: observations of the events; numerical models that allow us to simulate and predict them in the context of the larger climate within which they occur; and "theory." By theory, I mean our well-grounded and tested understanding of the first principles that govern the events and their relationship to climate, principles that can be expressed without resorting to numerical models. When observations, models and theory yield similar answers about how some type of event is related to climate, we become more confident in our understanding of the relationship. If one or more of the three is inadequate, or they are inconsistent with one another, we are much less confident.

Heat waves are the best example of a case where observations, models, and theory converge. Observations show heat waves increasing in frequency and intensity in most parts of the world; we understand well how heat waves are related to the climate in which they occur; and because climate models predict that they should be increasing in frequency and intensity. At this point in history, when a heat wave occurs, one can almost say with confidence that global warming made it *more likely*, *more intense* or *both*, even without doing a formal attribution study.

To take the other extreme, we know relatively little about how tornadoes are changing. The observations do show some changes in statistics of tornado occurrence – especially, increasing tendency for them to be bunched into large outbreaks, rather than spaced out more in smaller clusters. But the observations themselves are imperfect; and beyond that, we do not have the necessary theoretical understanding of how tornadoes are related to climate to be able to say with confidence that these changes are caused by warming, and climate models at this point provide only weak guidance.

Most kinds of extreme weather fall in between these extremes of understanding and ignorance.

We have good confidence that heavy rain events are increasing in many parts of the world: again observations, theory and models are all broadly consistent. Droughts and wildfires are both to some degree influenced directly by temperature, so we have good confidence that global warming increases either the frequency or intensity of these events under some conditions, although other factors that influence them may sometimes be more important.

In the case of wildfires such as those that have devastated the American west in recent years, warming and drying are the primary causes. Pre-historical evidence stored in tree rings and charcoal buried in lakes tell us that for thousands of years, periods of warming have coincided with periods of increased wildfire activity in this region. While fire is very complex and affected by much more than just climate, the data from recent decades indicate the same thing today as in the past: the hot years are the years with the most wildfire, and as temperatures have increased, burned areas have increased in step. It is entirely possible that on-the-ground human factors such as land management and accidental ignitions have set the stage for an especially potent fire response to warming in some areas, but the relationship between annual burned areas and temperature has nonetheless been strong and stable over the past few decades. We should plan for continued increases in western U.S. wildfire activity due to continued warming.

We have relatively little understanding of how winter storms are changing, except we know that warming makes some storms produce rain when they would have produced snow in the past (though when it remains cold enough to snow, warming can under some conditions increase the amount of snow).

In-Depth Example: Hurricanes

Of all types of extreme weather, hurricanes do the most damage. Hurricane Dorian's absolute devastation of the Bahamas is fresh in our minds; the U.S. was fortunate to escape major impacts from Dorian, but was not so fortunate with Hurricanes Michael and Florence last year, or Hurricanes Harvey, Irma, or Maria in 2017. Hurricanes are also the focus of my own research, and the issues around interpreting their relationship to climate are to some extent representative of those with other kinds of events. Hurricanes illustrate some broader issues around communicating and acting on our scientific understanding of the risk, as well as for their own intrinsic importance.

What We Know About Changes in Hurricanes

What do we know about how hurricanes are changing with climate? We can give the most precise answer if we break it down into different aspects.

The most certain way in which hurricane risk is increasing due to climate is that, because of sea level rise, coastal flooding due to hurricane storm surge is becoming worse. Storm surge occurs when the winds from a storm push the ocean onto the land. The total flooding is determined by the surge (the part produced by the wind), the tide, and the background average sea level. As sea level has risen – about a foot in New York City, for example, of which about eight inches is related to climate - for any given combination of storm and tides, the flooding is exacerbated by that amount. There is no doubt about this. The flooding Hurricane Sandy produced, for example, was due to nine feet of storm surge plus a high tide that was five feet above low tide. So the eight inches of additional water due to sea level rise was a small fraction of that, but still a significant one. There is no question this number will increase in the future; sea level rise projections are uncertain in magnitude, but certain in sign: sea level will only go up, not down.

Also the rain hurricanes produce is increasing. Rain-driven flooding from storms like Harvey and Florence is becoming exacerbated, perhaps somewhere between five and twenty percent per degree Celsius (or per 1.8 degrees Fahrenheit) of warming.

In the case of Harvey – and also Dorian in the Bahamas, though its damage was more due to wind and surge than rain – the disaster was made much worse by the slow forward motion of the storm, so that it stayed in one place for a long time. Several recent studies show that storms on average have been slowing down, and suggest that this is a consequence of climate change. These are relatively new findings, not fully understood or digested by the scientific community yet, so this conclusion is particularly uncertain. But the studies are of high quality, and their implications are very serious, so they should be taken into consideration as part of our overall assessment of risk.

Another fairly certain consequence of warming is that hurricane winds are strengthening. Again there is support for this conclusion from observations, theory, and numerical models. The evidence is particularly strong for the north Atlantic – the source of the hurricanes that threaten the United States. The magnitude of the increases in intensity we can attribute to warming is not clear; it may only be a few percent, but even that is significant. The strongest storms do by far the most damage, and increases in intensity at the high end mean more category four and five storms.

Because damage is proportional to wind speed cubed, or perhaps even a higher power, we see that for a given small percentage increase in wind, the damage increase is three or more times greater.

In contrast, some aspects of changes in hurricanes are almost entirely uncertain. In particular, we can say very little about how hurricane frequency – the total number of storms that occur each year – will change with warming. Because all other aspects of changes in hurricanes only matter if, where and when a hurricane occurs in the first place, this uncertainty about hurricane frequency limits our ability to assess overall hurricane risk in a changing climate.

We do not have a good understanding of what controls the overall number of tropical cyclones (tropical storm intensity and higher) on the earth presently, which is around 90 per year for the whole earth, around 11 for the Atlantic. Additionally, and we do not have any physical theory for how this should change as climate does. The observations lack any clear indication, mostly showing large fluctuations year to year and decade to decade that make it difficult to discern clear trends. Until recently, numerical simulations tended to show that hurricane frequency should decline with warming, but in the last few years simulations with a couple of the best models have instead produced increases. This produces a large uncertainty in our overall assessment of hurricane risk; if each storm on average produces stronger winds, heavier rains, and worse coastal flooding, but the total number of storms were to decrease enough, the total hazard - the probability of an event of a given magnitude at any given location – might still stay constant, or even decrease. But if the number of storms increases along with the intensities of their wind and rains, then we are in even bigger trouble.

With hurricanes we have the following complex situation: Some aspects are certainly becoming worse with warming; others are likely becoming worse, but with some uncertainty; and other aspects are very uncertain, such that changes in the overall hazard are also uncertain. This situation is broadly representative of other kinds of extreme weather events. The degree of uncertainty varies, but is usually substantial. Yet it would be a grave mistake to interpret this uncertainty as license to ignore the problem and postpone action on climate. There are at least two reasons for this.

Uncertainty Is Not Our Friend

Uncertainty about how the risk is changing means we have to accept some possibility of the worst outcome, namely that the risk is increasing at the upper bound of plausible scientific estimates. This is sometimes known as the "precautionary principle," and it is consistent with how human beings rationally deal with other kinds of risks in life, particularly when the worst outcomes would be truly serious.

Much of the uncertainty in our understanding of changes in extreme weather is due to the fact that our observational record is short while natural variability is large, so that it is difficult to separate the contribution of human influence from that natural variability. The climate fluctuates naturally from year to year, decade to decade, and even century to century. The gradual trends due to human-induced climate change are superimposed on these large fluctuations. With extreme events, the fluctuations are even larger because the events are - by definition – rare, so that the statistics are less conclusive.

To understand this, just flip a coin some number of times, and calculate the fraction of the time it comes up heads. Repeat with different numbers of coin flips, and notice that the more flips you have, the closer your average generally gets to 0.5. When we look at extreme events compared to regular weather, it's like having fewer coin flips. Now to understand the role of climate change, try to imagine that we are trying to determine whether the coin is fair, or whether the probability of heads has become, say, slightly greater than 0.5, though it was 0.5 in the past. This will be more difficult the fewer flips we have; that example is similar to the situation with hurricanes, since there are few of them compared to days with normal weather.

Further, climate scientists traditionally apply criteria for detecting and attributing trends that are very conservative: they are designed to minimize the risk of a so-called type 1 error (claim of a change when none is actually present, or "false alarm") but in doing so they maximize the probability of type 2 errors (failure to detect a change when one actually is present).

NOAA makes the public statement, as is currently visible on one of its web pages maintained at the Geophysical Fluid Dynamics Laboratory: "In the Atlantic, it is premature to conclude with high confidence that human activities–and particularly greenhouse gas emissions that cause global warming–have already had a detectable impact on hurricane activity." A few sentences later: ""Human activities may have already caused other changes in tropical cyclone activity that are not yet detectable due to the small magnitude of these changes compared to estimated natural variability, or due to observational limitations."

What NOAA is trying to say, in my view, is "there are changes, but we cannot show at 95% confidence that those changes could not have occurred in the absence of human-induced climate change." That may be true, but I would argue that that is not the right question to ask. We know that human-induced climate change is present.

The right question is: what is our best estimate of what the changes are, with what confidence? How wide is the range of possibilities that are reasonably consistent with the data, and what is the worst-case scenario?

The Most Important Thing to Understand Here Is: When It Comes to Disaster Risk, Uncertainty Is Not Our Friend

When faced with risks we can't assess precisely, but where we have some evidence that they may be increasing, choosing to ignore that evidence because of uncertainty is not prudent. Imagine you want to cross a highway. There are few cars on this highway, but they drive fast, and you can't see around a sharp corner. You don't know the probability that a car is coming, and none have come by for a for a while. Do you assume it's fine and walk across? Or, if there were an action you could take that would reduce your risk, even at some cost – say, walking to somewhere with a better view in both directions, even if it makes you late to where you need to be – wouldn't you do it?

Or, we can make the analogy a little closer using another risk that is hard to quantify: terrorism.

Imagine that a U.S. intelligence agency has some evidence that some bad people somewhere in the world may be planning an attack. The evidence is inconclusive, but strong enough to warrant concern. These bad people are having a meeting somewhere, and it is suspected that their agenda at that meeting is to plan the attack. U.S. agents are not present at the meeting, but have managed to plant a microphone in the room, connected to a transmitter so that they can hear the sound in the room at their offices in the U.S. But the room is noisy and the bad people are speaking quietly, so it is impossible to make out what they are saying, and thus impossible to be sure if they are really planning the attack or not. Would we want the U.S. agents to interpret this uncertainty as meaning everything is fine and no action needs to be taken? Or would we want them to take whatever measures they have at their disposal to prevent the attack, given whatever incomplete information they do have? In this analogy, the possible terrorist attack represents the possibility that hurricane frequency is increasing - along with hurricane wind intensity, rain, and coastal flooding – representing the greatest possible increase in risk. The "noise" is natural variability.

In this example, I think most of us would want to take action and the same is true, in my view, with respect to extreme weather and climate.

Changes in the Future Will Be Greater Than in the Past or Present

In addition to taking inappropriate comfort from uncertainty, another fallacy is to assess the human influence on extreme weather risk using only data from the present, while ignoring *the likely greater increases in the future*.

Human induced climate change has already caused changes in some kinds of extreme weather events. Attribution studies are now done in real time to assess to what extent any given event that just happened may have been influenced by global warming. These studies are important in helping the public to understand the links between climate and extreme weather, because they capture attention during the teachable moments right after major disasters.

But by focusing attention on the present, when the warming is less than it will be in the future, they can actually give the impression that climate change is less serious than it is, once we accept some responsibility to future generations. With many kinds of events – including hurricanes – event attribution studies give inconclusive results, because of the large natural variability and short, imperfect historical records (and sometimes also because numerical models are not quite good enough to do such studies). If an attribution study gives inconclusive results, as some do, that might leave the impression that climate is not changing that kind of event, and that this is one less reason for action now. Perhaps it would make more sense to wait until we see clearer indications of human influence in extreme events, and then take action. The problem here is that there is a long lag between action and result when it comes to greenhouse gas emissions. We need action now if we hope to reduce the impacts of climate change in the future.

The greenhouse gases already emitted by human activity have committed us to some additional warming beyond what has already been realized, due to the time it takes for the ocean to warm. We are almost certainly committed to additional warming beyond that due to the commitment baked into our current economic and energy systems – that is, absent much stronger and more immediate commitments to decarbonization than currently appear likely, greenhouse gas emissions will continue at sufficient rates to drive further warming for some time. If the climate were a ship, it would be a very large aircraft carrier or ocean liner – it can't be turned around quickly. As further warming proceeds, the changes in extreme weather will continue to grow.

We cannot wait until all the uncertainties have resolved themselves. To take the case of hurricanes, by the time we know with precision how much hurricane risk increases with each degree of warming, the risk will have increased quite a lot – that is how we will be sure, because only then will the data show it conclusively – and by then we will have baked in yet much more warming, warming that we could have avoided with earlier action.

Future Research Challenges

There are several different areas where additional research on extreme weather is urgently needed.

Short-term forecasts directly save lives and property. Weather forecasts, including those for extreme weather, have continuously improved from decade to decade since the mid-20th century. The three-day hurricane track forecast today, for example, is as good as the one-day forecast was 30 years ago – and two extra days of warning makes an enormous difference in emergency managers' abilities to save lives and property. This increase in forecast skill is an amazing success story of science and technology, even if the public doesn't always recognize it, and it has been largely driven by federal investment in research – much of which was authorized by this Committee. Such improvements will continue as long as the government sustains its support of the research into the observations, numerical models, data assimilation, and high-performance computing that form the backbone of the modern weather, water and climate enterprise – and the Congress continues to exercise constructive oversight on weather and climate research as it has done via the Weather Research and Forecasting Innovation Act of 2017.

On the longer time scale, an exciting development of the last decade or so has been the emergence of some skill in numerical models on the "subseasonal to seasonal" time scale – especially the subseasonal, meaning roughly 2-4 weeks ahead. This new capability is making it possible to produce forecasts of some phenomena on that time scale. But these forecasts remain mostly experimental and have only a very small amount of skill. The challenges are not just to figure out what can be usefully predicted and to make the predictions better, but also to figure out *how to communicate and use forecasts when they are only slightly better than no forecast at all.* You can make money if you bet on such a forecast over a long time, but much of the time it will still be wrong. Under what circumstances is such a forecast useful, and how can one make sure users understand its limitations and do not develop unrealistic expectations that are sure to be disappointed? For example, there might be moves that could be taken to begin pre-positioning people or materials well in advance of a wildfire or hurricane that appears possible in two weeks, due to a subseasonal forecast, such that the response to a disaster later will be more effective, but that are sufficiently inexpensive that there will be little regret if the event does not materialize.

For more accurate detection and attribution of changes in extreme weather events due to human-induced climate change several things are needed. First, as in all climate research, *the observational network must be sustained over time*, or better, strengthened, so that we can maintain the long-term records that are necessary to document climate change, including its manifestation in extreme events. Second, *climate models need to be continuously improved* – the US should maintain and build on its strength in climate modeling. Third, and perhaps least appreciated but equally important, *fundamental understanding of the relationship between climate and extreme weather events must be improved*. This is essential to our confidence in our interpretations of the observations and the models. In the case of hurricanes, we lack a plausible candidate theory that might explain the number of hurricanes on the planet each year and how that should change. This makes us totally reliant on numerical models which, though rapidly improving, are still not adequate to answer the question on their own. The Federal agencies that support climate research should more explicitly prioritize work whose goal is to achieve such basic understanding, as much as it prioritizes work whose goals are to improve models or observations.

Perhaps most urgently needed, though, in my view, is *research that quantifies the risks from extreme weather, and their changes as the climate warms, in terms of their impacts on human society: economic losses, fatalities and human health impacts, harm to ecosystems, etc.*

For most of the past decade, I have been interacting closely with colleagues in the insurance industry. They use tools called "*catastrophe models*" to assess the risks from extreme weather events. These industry catastrophe models are designed to solve the problem that most disaster losses come from a few large events, those events are rare, and often modern recorded history has no analog. Before Hurricane Sandy, the last comparable event in New York City occurred in 1821. There were neither good measurements, nor did the city have anywhere near its size or population in 2012. Understandably, the impacts were not comparable. How could one have assessed the risk, pre-Sandy, lacking good, recent historical analogs? Catastrophe models generate large numbers of synthetic events – virtual storms, say, that are realistic, but fill such gaps in history. The models calculate not only the storms' geophysical dimensions but also the impacts they would have on buildings and infrastructure. Essentially, catastrophe models produce synthetic histories from which more representative estimates of risk can be produced.

Such catastrophe models have served the insurance industry well until now, but they have significant limitations in the new environment we face today, where climate change is an established fact and the industry, along with most of the rest of the private and public sector, needs to understand how extreme weather risk is changing. Because of the way catastrophe models have been developed, based closely on historical data, they implicitly assume that the present and near future will be similar to the past, and thus do not adequately capture climate change. The most influential and widely used models are also proprietary, meaning the details of their construction and output are not subject to open scientific debate and peer review. Finally, they are designed to be used in places and for assets where the insurance industry has significant exposure, but tend to be less accurate or nonexistent elsewhere. Thus for calculating property damage risk due to hurricanes in the U.S. they may be pretty good, while they basically can't be used to calculate human health risks from hurricanes in Mozambique, for example.

Some public, academic and nonprofit catastrophe models do exist, but these largely share the weaknesses of the private sector ones, or have even more severe limitations in some respects. The science of climate-aware catastrophe modeling is in its infancy. Yet with the rapidly increasing pressure for climate-related financial disclosure in the private sector, and for increased resilience and adaptation to extreme weather risk at the state, local and federal level in the public sector, there is a rapidly growing need to overcome these limitations.

It is time for the Federal science agencies to invest in a set of open - source tools to assess changing extreme weather risk in a way that is practically useful for real decisions, accounts for climate change, and whose methodologies and assumptions can be debated openly, in the peer-reviewed literature and elsewhere. This will highlight their strengths, weaknesses, and appropriate uses – including the best way to quantify the uncertainties, which will be even larger when climate change is accurately integrated into models and forecasts. They would be available for the insurance industry, and the rest of the private sector, to use (alongside the existing proprietary models, which should and would remain in place) but also be available for use by the governments, such as to inform cost-benefit calculations for building physical flood defenses or any other measure being designed to increase resilience.

The private sector would benefit greatly from the existence of such tools, but for them to be trusted it is important that no company or other private interest "own" them, thus the funding needs to come from the government, or perhaps a public-private partnership. A federal research program in such a direction could be guided, for example, by a steering group with representatives from the private sector, government, and academia.

Concluding Remarks

Thank you for the opportunity to participate in today's hearing. I also want to thank this Committee and your colleagues on both sides of the aisle and both sides of the Capitol for your steadfast support for the Nation's research enterprise. I suspect you have many difficult decisions to make on where to allocate the public's resources. Your support for research and education has helped this Nation main tain its competitive edge and allowed science to contribute to the nation's national, economic and environmental security. I would be pleased to answer any questions or provide additional follow up information that may be useful to the Committee.

Dr. Adam H. Sobel, Professor, Applied Physics & Applied Mathematics and Earth & Environmental Sciences, Columbia University, New York, NY

Dr. Adam Sobel is a professor at Columbia University's Lamont-Doherty Earth Observatory and Fu Foundation School of Engineering and Applied Sciences. He is an atmospheric scientist who specializes in the dynamics of climate and weather, particularly in the tropics, on time scales of days to decades.

A major focus of his current research is extreme events - such as hurricanes, tornadoes, floods, and droughts, and the risks these pose to human society in the present and future climate. He leads the Columbia University Initiative on Extreme Weather and Climate. Together with colleagues in both academia and the insurance industry, Sobel has also been developing models to assess the risk of rare but extremely damaging extreme weather events, particularly tropical cyclones, tornadoes, and hail.

Sobel holds a Bachelor's degree in Physics and Music from Wesleyan University, and a Ph.D. in Meteorology from the Massachusetts Institute of Technology. In the last few years, he has received the Meisinger Award from the American Meteorological Society, the Excellence in Mentoring Award from the Lamont-Doherty Earth Observatory of Columbia University, an AXA Award in Extreme Weather and Climate from the AXA Research Fund, and an Ascent Award from the Atmospheric Sciences Section of the American Geophysical Union. Sobel is author or co-author of over 125 peer-reviewed scientific articles, and his book about Hurricane Sandy, *Storm Surge*, published in October 2014 by Harper-Collins, received the 2014 Atmospheric Science Librarians International Choice Award in the popular category and the 2016 Louis J. Battan Award from the American Meteorological Society.

September 2019

Chairwoman JOHNSON. Thank you. Dr. Moore.

Testimony of Dr. Berrien Moore, Director, National Weather Center, University of Oklahoma

Dr. MOORE. Thank you, Congresswoman Chairwoman Johnson, Ranking Minority Member Lucas. I'm delighted to be here. I'm Berrien Moore, Director of the National Weather Center and Dean of the College of Atmospheric and Geographic Sciences at the University of Oklahoma. My testimony is mine alone and not representative of those organizations.

Having listened carefully to my colleagues, I've adjusted my verbal remarks somewhat to focus on areas that they did not touch upon but that certainly does not mean that I'm in disagreement with them. Regarding climate and weather, one may think of it this way is that one of the great challenges is that some of the statistical properties that we've historically relied upon to help us in our forecasting are being changed.

There are parts of Oklahoma that you can drive through by just simply looking in the rearview mirror because it is very flat. We are now entering the Rocky Mountains, and looking in the rearview mirror may not serve us well. So, a consequence, the numerical weather prediction models are those models that allow us to look forward into time and help us drive the automobile better.

Unfortunately, as we all know, and Congress has spoken about this through the *Weather Innovation Act*, as well as the *Drought Act* of last year, bringing into existing the Earth Prediction Innovation Center that we as a country are not doing as well as some of our European colleagues. This is particularly unfortunate given the fact that we as a country seem to be subject to more weather extremes than many other parts of the world. And therefore, that failure of us as a country is particularly painful.

I'd like to now direct my testimony on three areas. What new observations might we need to improve the situation? And then how can we assimilate those observations better in our numerical weather prediction models to improve our predictive capability? And finally, how do we improve the models themselves?

Regarding observations, I think there are three principal topics. First of all, we have a very important weather radar system for the United States, but it is an aging radar system. The service life expectance has allowed us to extend the life of these radars, but we are going to need a new weather radar system for the United States certainly by 2040. Well, the implementation of that would certainly take 5 years, the procurement of that would certainly take 5 years, so now we're at 2030. Well, we have to have the requirements and the technological base for this new weather radar system by certainly 2028. That's just 8 years away. We need to get on with this.

Second, satellites are extremely important to us in terms of our weather prediction, the GOES system, the JPSS system, but what I see is missing is we do not have a sounder, something that tells us about the humidity and the temperature and the water vapor in the lower part of the atmosphere. I think what we need to do is to fly a hyperspectral environmental sounder. And what we might do is put it on a commercial communications satellite as a hosted payload. This is something that we are exploring in the country. In fact, three of NASA's upcoming missions are going to be via hosted payload.

Third—and Congressman Lucas spoke about this—the national mesonet, this is the gold standard of national mesonets is in the State of Oklahoma. Every State should have a gold standard. Every State merits the observational network that we in Oklahoma enjoy. And as we look to the future, one of the problems for the national mesonet is it's a ground-based system. As I just said with a hyperspectral sounder, we need a space-based system, but it needs to be complemented by a ground-based system that allow us to look into the third dimension. Certainly with the increase of drones there ought to be a way to do this, but it's going to require foresight and action by the Congress.

Finally, I think that EPIC really does offer us an opportunity, the Earth Prediction Innovation Center, to go forward and improve the data assimilation and modeling by allowing us to marshal the full scientific enterprise of the United States to move on this problem. Thank you. I'm honored to be here.

[The prepared statement of Dr. Moore follows:]

TESTIMONY OF DR. BERRIEN MOORE, III, DEAN OF THE COLLEGE OF ATMOSPHERIC & GEOGRAPHIC SCIENCES, CHESAPEAKE ENERGY CORPORATION CHAIR IN CLIMATE STUDIES, DIRECTOR OF THE NATIONAL WEATHER CENTER

Chairwoman Johnson, Ranking Member Lucas, Members of the Committee, I am Berrien Moore, Director of the National Weather Center at the University of Oklahoma. Thank you for the chance to testify at today's hearing on severe weather events, a forum that is timely in its relevance as we are now in the middle of the nation's annual hurricane season.

Apart from these often time catastrophic events, many decisions in important sectors of Oklahoma's and the Nation's economy, such as agriculture, energy, and transportation, depend on weather information. About one-third of the U.S.'s $18 trillion gross national product is generated by industries vulnerable to changes in the weather. Weather and climate information affect every aspect of life and severe weather events lead to serious disruptions with both short-term and long- term impacts on the economy, public safety, and health.

There were 14 severe weather events that caused at least $1 billion in damages in the U.S. in 2018, according to the National Centers for Environmental Information. Dramatic events such as tornado out breaks such as the ones in Moore, Oklahoma (1999 and 2013), or landfalling hurricanes like Harvey in Houston (2017), Florence in the Carolinas (2018), and Michael in Florida (2018), and Dorian earlier this month as well as the recent tragic wild fires in California all highlight for the general body politic the vulnerability of our society to weather.

The fact that climate is changing adds to the challenge of weather prediction; looking at weather histories is, in a sense, less useful. While it is very difficult if not impossible to attribute a specific weather event to the fact that climate is changing, we can state that the statistical properties of the weather system are changing; stating it more scientifically, the weather system is no longer a relatively stationary process. This fact alone makes the challenge of weather forecasting more difficult and meeting the challenge more important.

We at the National Weather Center understand well and support firmly the National Oceanic and Atmospheric Administration (NOAA) in carrying out its Mission of protecting lives, property and the economy.

We also were greatly appreciative of the Congress for passing the Weather Research and Forecasting Innovation Act of 2017.[66] This bill, which was the *first* major weather authorization legislation to become law in over two decades, requires NOAA to conduct a two-pronged program; first, to realize "improved understanding of forecast capabilities for atmospheric events and their impacts, with priority given to the development of more accurate, timely, and effective warnings and forecasts of high impact weather events that endanger life and property;" and second, to "collect and utilize information to make reliable and timely foundational forecasts of subseasonal and seasonal temperature and precipitation." Predictions at these time scales are critical for many practical applications, such as water management (drought vs. pluvial years), prevention of vector borne diseases like malaria or west Nile virus, energy management (heating and cooling needs), and especially agriculture.

The leading numerical weather prediction (NWP) models, including the NOAA Global Forecast System (GFS), have exhibited steadily increasing skill over the past 15 years. However, the European Centre for Medium-range Weather Forecasts (ECMWF) and the UK Meteorological Office (UKMO) consistently outperform the NOAA's Global Forecast System. This fact was a catalyst in the creation of the Weather Innovation

---

[66] The Bill noted that *Subseasonal forecasting is forecasting weather between two weeks and three months and seasonal forecasting is between three months and two years.* https://www.washingtonpost.com/news/capital-weather-gang/wp/2017/04/04/congress-passes-comprehensive-weather-forecasting-and-research-bill/?noredirect=on&ut:m_term=. bfe4ee12ce59 https://docs.house.gov/billsthi sweek/20170403/HR353.pdf.

Act of 2017. At its core it recognized that improvements in weather forecasting can come from improvements in

- measurements (Observations),
- ways to ingest these measurements (Assimilation), and
- physical representation of weather processes (Modeling).

Importantly, this Act serves as a guidepost for providing NOAA with the resources to make it happen.

More recently, the National Integrated Drought Information System Reauthorization Act of2018 (NIDISRA), instructed NOAA to:

- prioritize improving weather data, modeling, computing, forecasting and warnings; and
- establish the Earth Prediction Innovation Center (EPIC) to accelerate community-developed scientific and technological enhancements for numerical weather prediction.

My testimony will now focus on certain specifics that give increased definition to the three foundational elements in numerical weather prediction:

- measurements (Observations),
- ways to ingest these measurements (Assimilation), and
- physical representation of weather processes (Modeling).

Observations

It is my opinion that there are three critical Observational Areas: a) The Next Generation Phased Array Weather Radars that are needed to replace the current WSR-88D radar system; b) Hyperspectral sounding from geostationary orbit to provide vertical profiles of temperature, hun1idity, cloud condensate over CONUS, and c) a three dimensional

National Mesonet to provide even greater insight into the planetary boundary layer.

The Next Generation Phased Array Weather Radars must be dual polarized and all-digital in order to discriminate cloud liquid water from ice and to realize the needed fast scanning that is critical in severe weather situations.

Given the age, costs and capabilities of the current WSR-88D radar system; given a service life extension program (SLEP) for the WSR-88D system, and given the importance of weather and weather forecasting for the country, it seems prudent to establish, fund, and thereby implement a plan to realizing the next generation meteorological phased array radar network. Having completed the extensive service life extension program for the WSR-88D, NWS estimates that it can be reliably operated until about 2040. Estimating that five years will be required for deployment of the new radar network, m1d allowing for the preceding acquisition effort - requirements publication, contract award, first article development and operational test- it will be necessary to establish validated requirements and the technological base for Next Generation Phased Array Weather Radars by 2028.

The needed research and development, which has its beginning in the Multi-function Phased Array Radar *(MPAR)* project, must be accelerated and focused on the weather-relevant capabilities and requirements. This is central m1d needs to begin now - with accelerated work commencing in Fiscal Year 2020.

With respect to the collection of hyperspectral data, I believe that NOAA should be given authority to proceed toward an on-orbit demonstration of a hyperspectral instrument at geostationary orbit as a hosted payload. I note that the Weather Innovation Act of 2017:

- requires NOAA to "establish a tornado warning improvement and extension program" that can increase the advanced warning of tornadoes to beyond one hour. (Section 103);
- details Congressional interest in assessing the value of data. . . "from a geostationary hyperspectral sounder." (Section I 07), and

- pe1mits "placement of weather satellite instruments on co-hosted government or private payloads."

There are multiple reasons why hyperspectral sounding data is needed:

- Time-sensitive information on water vapor transportation both horizontally and vertically (especially in the boundary layer) is key to forecasting severe storms, including deep convective, tornadic storms over the continental US (CONUS).
- Moisture and wind observations are critical for storm prediction through assimilating into regional or storm scale numerical weather prediction models over CONUS.
- Improving the prediction of hurricane intensity and landfall position requires detailed and continuous information on vertical wind profiles.
- Increased destruction and damage from severe storms such as hurricanes and tornadoes are gaining broader public awareness and have heightened public desire to improve our forecasting infrastructure.

Only an advanced hyperspectral sounding instrument from geostationary orbit can provide the needed detailed moisture and dynamical motion information with high temporal and high spatial resolution into the boundary layer. Advanced IR sounders, such as the Atmospheric Infrared Sounder (AIRS) on Aqua and the Cross-track Infrared Sounder (CriS) on Suomi-NPP (SNPP) and the Joint Polar Satellite System (JPSS) series, do provide high vertical resolution atmospheric sounding information that are already improving the forecast skill in numerical weather prediction models. However, those high-quality measurements are only available in sun-synchronous LEO at a revisit rate of every 8 hours. For ground-breaking severe weather and near-term forecasting of threatening weather in the US, near real-time soundings are necessary to detect changes in the atmosphere as they occur, and this requires a geostationary orbit over CONUS. In the case of rapidly forming tornadoes, minutes instead of hours could mean the difference between life and death.

There is an emerging and exciting pathway to getting instrumentation into geostationary orbit through a commercial communication geostationary satellite as a *host.* The US Air Force was the first to use this approach for their "Commercially Hosted Infrared Payload (CHIRP) and more recently NASA flew the GOLD (Global-scale Observations of the Limb and Disk (GOLD) instrument as a hosted payload in GEO. In fact, NASA intends to fly three Earth Venture Instruments/Mission as hosted payloads in geostationary orbit: the Tropospheric Emissions: Monitoring of Pollution *(TEMPO),* the Geostationary Carbon Observatory (GeoCarb), and the Geosynchronous Littoral Imaging and Monitoring Radiometer (GLIMR).

There is the same possibility for getting a hyperspectral sounder to geostationary orbit between GOES East and GOES West. Combining a host satellite with a l-IES-like Prototype Instrument payload as a FY 2021 budget initiative would be a "Path Forward" for the US to have a hyperspectral sounder in geostationary orbit much sooner than would otherwise be possible. This would be consistent with the recent NOAA EPIC initiative which seeks both improved measurements and data assimilation into the weather models and is responsive to the Weather Research and Forecasting Innovation Act of 2017.

With respect to the last element of observations, the National Mesonet, it is essential for this Committee to accentuate its strong support for this program and its expansion to provide further innovative means to collect additional data-particularly from the boundary layer. The National Mesonet program has enjoyed strong Congressional support but NOAA budget requests for it continually propose large reductions despite its high value and importance for aiding in improving forecasts for severe weather. A formal authorization of the program by this Committee would be welcome in this effort.

The deployment of small unmanned aircraft systems (UAS) to collect operationally 24-7 *in situ* vertical profiles of the the1modynat1lic and kinematic state of the atmosphere in conjunction with other weather observations from the National Mesonet (https://nationalmesonet.us) could significantly improve weather forecasting skill and resolution. High-resolution vertical measurements of pressure, temperature, humidity, wind speed at1d wind direction are critical to the understanding of atmospheric boundary layer processes and directly support and complement the sounding from a hyperspectral sounder discussed above. I believe that the use of UAS will expat1d rapidly over the coming decade, and the weather community needs to be in a position to exploit that expansion and thereby create a 3D Mesonet.

The Oklahoma Mesonet is of extraordinary value to Oklahoma and surrounding states and this value will be dramatically enhanced by its extension into the vertical dimension. NOAA through its Office for Oceanic and Atmospheric Research needs to expand its UAS progran1dramatically and thereby establish the foundation for an operational network could be realized to better characterize the atmospheric boundary layer, to identify threats of severe weather, and to improve fundamentally forecast of severe weather. A useful summary of the value of such a 3-D Mesonet was recently characterized by a number of my colleagues at the University of Oklahoma that was published in a recent edition of the journal *Sensors* that was dedicated entirely to the use of wm1rumed aircraft for atmospheric science investigations.[67]

Finally, regarding Observations, the National Weather Center (NWC) and the Radar· Innovations Laboratory (RIL) at the University of Oklahoma are in a leadership position in developing a new generation of all digital, phased-array polarimetric radars. The NWC and RIL house together about 800 people, which includes research scientists, operational meteorologists and climatologists, engineers, technicians, support staff, and graduate students. The researchers based at the NWC are working on a wide variety of projects including radar meteorology, mesoscale meteorology, cloud physics and lightning, synoptic meteorology, ru1d climate variability and change. In addition, within the NWC, the Center for

---

[67] Chilson et al., *Moving towards a Network o(Autonomous VAS Atmospheric Profiling Stations for Observations in the Earth's Lower Atmosphere: The 3D Mesonet Concept, Sensors* 2019, /9(12), 2720.

Autonomous Sensing and Sampling, in collaboration with NOAA's National Severe Storms Laboratory, is developing advanced drone-based weather observations. The Oklahoma Climatological Survey (OCS) in the NWC was established by the State Legislature in 1980 to provide climate services to the people of Oklahoma. The Survey maintains an extensive array of climatological information, operates the Oklahoma Mesonet, which is the world leader in ground-based weather observation, and OCS hosts a wide variety of educational outreach and scientific research projects. Finally, given OU's role as the lead institution on GeoCarb, we understand well the challenges and opportunities of hosted payloads as well as the scientific issues associated with the role of the boundary layer in severe weather. We stand ready to help NOAA in the arena of Observations.

Assimilation and Modeling

It is most appropriate to discuss data assimilation into models in the same text as we discuss modeling *per se.* As noted above the European Centre for Medium-range Weather Forecasts (ECMWF) and the UK Meteorological Office (UKMO) consistently outperform the NOAA's Global Forecast System. This fact led, in part, to both the Weather Innovation Act of 2017 and the National Integrated Drought Information System Reauthorization Act of 2018 (NIDISRA); the latter instructed NOAA to establish the Earth Prediction Innovation Center (EPIC) to accelerate community-developed scientific and technological enhancements for Numerical Weather Prediction.

Two important events have followed: development of a Working Vision Paper drafted by NOAA Scientists[68] that sets a road map for EPIC and a subsequent Community Workshop in early August 2019. Both the Vision Paper and the Workshop recognized that at the heart of EPIC is the need for a Unified Forecast System (UFS) which would be an end-to-end

[68] Cikanek et al., A Vision Paper for the Earth Prediction Resource Center, Version 5.0, May 28, 2019.

system for Earth System Prediction that encompasses the assimilation of observations, development of initial conditions, execution of the prediction model, post-processing and analysis of model output, verification of results, and the delivery of products. Obviously, there must be a software infrastructure to support this entire process.

The "prediction model" will be structured around a dynamic core (the Finite-Volume on a Cubed-Sphere or FV3) that was developed by NOAA's Geophysical Fluid Dynamics Laboratory in Princeton. The FV3 core brings a new level of accuracy and numeric efficiency to the model's representation of atmospheric processes such as air motions and was selected in 2017 to be the foundation for the next generation weather models. This capability is at the center of improvements in NWP.

NOAA needs the resources and "encouragement" to move forward with the EPIC Initiative sooner than later. Moving ahead with the EPIC Initiative is of first order importance.

The National Weather Center stands ready to help. The NWC is developing new techniques and novel methodologies for data assimilation and ensemble prediction. Researchers at NWC are applying these techniques from convective scale to global scale in modeling systems to realize a step-change in predictive skill. Importantly, there is a commitment and capability to transition this research and development into operations. Further, the NWC is in a leading position in development of improved convection resolving physics and applications of NOAA's new FV3 dynamic core.

In closing, let me return to the Weather Research and Forecasting Innovation Act of 2017. As the Committee knows well, tills Act requires NOAA's Office of Oceanic and Atmospheric Research (OAR) to conduct a program aimed at developing *"an improved understanding of forecast capabilities for atmospheric events and their impacts, with priority given to the development of more accurate, timely, and effective warnings and forecasts of high impact weather events that endanger life and property."* Certainly, the Earth Prediction Innovation Center, which is to be focused on "advancing weather modeling skill, reclaiming and maintaining international leadership in the area of numerical weather prediction, and improving the transition of research into operations" directly address the first charge of the Weather Innovation Act, which deals with high impact weather events. Given that the subject of the Hearing is on Extreme Weather, my testin1ony has focused more on the first charge in the Innovation Act.

However, the Act also asks for the National Weather Service to *"collect and utilize information to make reliable and timely foundational forecasts of subseasonal and seasonal temperature and precipitation. Subseasonal forecasting is forecasting weather between two weeks and three months and seasonal forecasting is between three months and two years."* I think that Weather Prediction on Subseasonal to Seasonal Time Scales, which sits beyond the range of classic weather prediction models (less than 10 days) but far shorter than the time scale of climate models (several decades) is an emerging grand challenge in weather/climate prediction.

Given the weight of all these challenges, it seems appropriate for this Committee and the Congress at large to consider the creation of an external, peer reviewed decadal survey process for weather - distinct from that for Earth science because of the more operational nature of weather data.

I would encourage the Committee to have additional hearings on this important challenge.

I appreciate the opportunity to share my views with the Committee and I look forward to our discussion.

College of Atmospheric & Geographic Sciences
ags.ou.edu

National Weather Center
nwcnorman.org

The University of Oklahoma
www.ou.edu
College of Atmospheric & Geographic Sciences

Dr. Berrien Moore
Dean, College of Atmospheric & Geographic Sciences
Chesapeake Energy Corporation Chair in
Climate Studies
Director, National Weather Center

Dr. Berrien Moore III is an internationally recognized Earth scientist who has been honored by National Aeronautics and Space Administration (NASA) and the National Oceanic and Atmospheric Administration (NOAA). He received his Bachelor of Science in Mathematics in 1963 from the University of North Carolina and his PhD in Mathematics in 1969 from the University of Virginia.

Berrien Moore III joined the University of New Hampshire (UNH) mathematics faculty in 1969 and became a tenured professor in 1976. He was recognized by UNH in 1992 for research excellence and was named University Distinguished Professor in 1997. From 1987 to 2008, Moore served for as the Director of the Institute for the Study of Earth, Oceans and Space at UNH. During his period at UNH, Moore also held numerous visiting scientist positions including visiting Senior Scientist at the Laboratoire de Physique et Chimie Marines at the Universite Pierre et Marie Curie in Paris and at the Institute of Meteorology at the University of Stockholm. Earlier, he served as a Senior Research Fellow at the East-West Center in Honolulu and as a Fellow at the Marine Policy and Ocean Management Program at the Woods Hole Oceanographic Institution.

In February 2008, Moore left UNH to serve as the founding Executive Director of Climate Central, a think-tank based in Princeton, New Jersey, which is dedicated to providing objective and understandable information about climate change.

In the summer of 2010, Moore joined the University of Oklahoma, where he holds the Chesapeake Energy Corporation Chair in Climate Studies. He also serves as Dean of the College of Atmospheric and Geographic Sciences, Director of the National Weather Center, and Vice President for Weather and Climate Programs.

He has published extensively on the global carbon cycle, biogeochemistry, remote sensing, environmental and space policy, and mathematics. In December 2016, NASA selected Geostationary Carbon Cycle Observatory (GeoCarb), as the Second Earth Venture Mission. Professor Moore is the architect and Principal Investigator on GeoCarb Mission. (https://www.nasa.gov/press-release/nasa-announces-first-geo-stationary-vegetation-atmospheric-carbon-mission)

Professor Moore chaired the overarching Scientific Committee of the International Geosphere-Biosphere Programme (IGBP) from 1998 to 2002 and as such led the July 2001 Open Science Conference on Global Change in Amsterdam and is one of the four architects of the Amsterdam Declaration on Global Change.

He was the Coordinating Lead Author for the final chapter, "Advancing our Understanding," of the Third Assessment Report of the Intergovernmental Panel on Climate Change (IPCC), which was released in Spring 2001, and as such has been honored for contributing to the award of the 2007 Nobel Peace Prize to the IPCC.

He has contributed actively to committees at the National Academy of Science; over the last two decades, he served as Chairman of the National Research Council's (NRC's) Committee on International Space Programs of the Space Studies Board that, in collaboration with the European Space Sciences Committee, jointly published *U.S.-Europe Collaboration in Space Science*. In 1999, he completed his Chairmanship of the National Academy's Committee on Global Change Research with the publication of *Global Environmental Change: Research Pathways for the Next Decade*.

From 2004 to 2007, Moore co-chaired (with Rick Anthes, Past-President of University Cooperation of Atmospheric Research) the NRC decadal survey, *Earth Observations from Space: A Community Assessment and Strategy for the Future*. This was the first Decadal Study for the Earth Sciences and had significant impact in the United States. More recently, he was member of the Informing Panel of the overall NRC study on *America's Climate Choices* and was a member of the Committee that issued the 2013 report *Beyond Landsat and Beyond: Meeting the Needs for Sustained Land Imaging*.

From 1984 to 1988, Professor Moore served as a committee member of NASA's senior science advisory committee, the Space and Earth Science Advisory Committee (SESAC), which published its report in 1986: *The Crisis in Space and Earth Science: A Time for a New Commitment*. In 1988, he assumed Chair of SESAC and served until 1992. He concurrently served as a member of the NASA Advisory Council.

Additionally, he has served NASA through a variety of other positions including Chair of the EOS Payload Advisory Committee, member and Chair of NASA's Earth Science and Applications Committee, Goddard Space Flight Center Review Committee, and member of the Executive Committee for the Earth System Science Study. He currently serves on the JPL Advisory Council.

From 2004-2005, Moore chaired the NOAA Research Review Team and has served on the Research and Development Portfolio Review Team for the NOAA Science Advisory Board.

In recognition of his lifelong dedication to science, he has been the recipient of numerous honors, including the NASA Distinguished Public Service Medal for outstanding service to the agency – NASA's highest civilian honor; the NOAA Administrator's Recognition Award; and the 2007 Dryden Lectureship in Research Medal by the American Institute of Aeronautics and Astronautics. In 2017, Moore received the Distinguished Alumnus Award from the University of North Carolina. He is an elected Fellow of the American Meteorological Society and International Academy of Astronautics.

Chairwoman JOHNSON. Thank you very much. Dr. Bostrom.

TESTIMONY OF DR. ANN BOSTROM, WEYERHAEUSER ENDOWED PROFESSOR IN ENVIRONMENTAL POLICY, UNIVERSITY OF WASHINGTON

Dr. BOSTROM. Good morning, Honorable Chairwoman Johnson, Ranking Member Frank Lucas, and Members of the Committee. Thank you for your invitation to speak on the urgent matter of extreme weather events and climate change. I'm Ann Bostrom, Weyerhaeuser Endowed Professor of Environmental Policy at the Evans School of Public Policy at the University of Washington. I teach research methods, decisionmaking, and environmental policy courses, and I work to increase interdisciplinary research on the environment to bridge science and society and to ensure that investments in basic sciences are also benefiting our communities.

I've also contributed to National Academies' reports, including "Communicating Science Effectively: A Research Agenda," and I co-chaired with the eminent William Hooke, the National Academies' 2018 report "Integrating Social and Behavioral Sciences within the Weather Enterprise."

In addition to climate change and extreme weather events, my research investigates other hazards and the perception of communication of what we know and can do about such hazards, as well as scientific and management uncertainties.

Thirty years ago in my first studies of climate change risk perception, communication, and decisionmaking, scientists and laypeople voiced their expectations of more extreme weather as CO_2 emissions from our fossil fuel use warmed the planet. Now the scientific evidence is overwhelming that climate change has contributed to extreme weather events in recent years, increasing their severity.

Despite the phenomenally improved forecasts that government research investments have enabled over recent decades, we have failed to forestall catastrophic damages to many of our communities from hurricanes, floods, droughts, and wildfires. To protect lives and property and to realize the full value of the investments made in the physical sciences, we need to invest in social and behavioral sciences of extreme weather and climate change. People need to know what to do when a tropical storm or hurricane threatens, for example, how driving might be affected and how to evacuate.

People intuitively understand that there are uncertainties in weather forecasting. They do not, however, always interpret visual and other forecast uncertainty information in the way that forecasters and emergency managers wish or expect. People also tend to be more prepared for an event when they have prior experience of it.

But while a plurality of people in the U.S. have long thought climate change will lead to more extreme weather events, their experiences may not be predictive of the weather extremities climate change will bring. Tropical Depression Imelda dumped 3 feet of rain in 24 hours, which caught people by surprise, despite Harvey. Much remains to learn about how best to communicate forecasts and forecast uncertainties in these circumstances.

The careful experimental research that has been conducted to date shows that people can make better decisions if they are given explicit uncertainty information based on the best scientific forecasts and tailored to their decision context. But there is a very large need for additional research on communicating uncertainty for different decision contexts, research that brings social, behavioral, and other sciences together to determine how climate and weather information can most effectively be integrated, analyzed, and delivered to help forecasters, emergency managers, drivers, indeed all of us make better decisions.

The National Science Foundation, the National Oceanic and Atmospheric Administration, and other agencies appear to be increasing their investments in such research. But to date they constitute only a very small proportion on the order of less than 10 percent of the weather-related research investments. These investments have funded pilot programs like the CASA Dallas-Fort Worth Living Lab Program, which provides timely, tailored, human-scale forecasts on personal devices and surveys users to achieve continuous improvement.

To fully realize these and other lifesaving advances on a national and international scale will require scientific leadership and capacity-building across the public and private sectors as well.

Thank you for the opportunity to speak with you today on these critical issues relating to extreme weather. I look forward to your questions.

[The prepared statement of Dr. Bostrom follows:]

WRITTEN TESTIMONY OF ANN BOSTROM, WEYERHAEUSER ENDOWED PROFESSOR OF ENVIRONMENTAL POLICY, DANIEL J. EVANS SCHOOL OF PUBLIC POLICY AND GOVERNANCE, UNIVERSITY OF WASHINGTON

Thank you Chairwoman Johnson and Ranking Member Lucas for your invitation to testify on the urgent matter of extreme weather events and climate change. I am Ann Bostrom, Weyerhaeuser Endowed Professor of Environmental Policy in the Daniel J. Evans School of Public Policy at the University of Washington. I study risk perception, communication, and decision making under uncertainty in applied contexts like climate change and extreme weather, usually with interdisciplinary teams. In addition to climate change and extreme weather events, my research investigates other hazards, and the perception and communication of what we know and can do about the risks they pose, as well as scientific and management uncertainties. I also teach research methods, decision making, and environmental policy, with the aim of informing and improving the analysis and management of environmental and health risks. Achieving this requires advances in social and behavioral sciences along with advances in other sciences, and bridging science and society to ensure that investments in basic sciences are benefitting our communities. I have also contributed to National Academies reports that pertain to this hearing, including *Communicating Science Effectively: A research agenda*,[69] and I had the pleasure of co-chairing with the eminent William Hooke the National Academies 2018 report *Integrating Social and Behavioral Sciences within the Weather Enterprise*.[70]

Changing Perceptions, Yet Still Unanticipated Extremes

Thirty years ago in my first studies of climate change risk perception, communication and decision making, scientists and lay people voiced their expectations of more extreme weather as CO_2 emissions from our fossil fuel use warm the planet.[71] Our survey respondents in the early 1990's thought that a rise in mean sea level from climate change would increase the severity of storm surge incursions into coastal areas, but they did not anticipate that New York might flood.[72] When we replicated this survey a decade ago (2009), almost half (42%) of our respondents though it was true or possibly true that global warming would cause the ocean to flood all of the city of New York.[73] Seven years ago this October, superstorm Sandy flooded New York streets and subways, with climate change at least partly to blame.[74] The scientific evidence is now overwhelming that anthropogenic climate change has contributed to extreme weather events in

---

[69] National Academies of Sciences, Engineering, and Medicine. 2017. Communicating Science Effectively: A Research Agenda. Washington, DC: The National Academies Press. doi: https://doi.org/10.17226/23674.

[70] National Academies of Sciences, Engineering, and Medicine. 2018. Integrating Social and Behavioral Sciences within the Weather Enterprise. Washington, DC: The National Academies Press. https://doi.org/10.17226/24865.

[71] Bostrom, A., Morgan, M.G., Fischhoff, B. and Read, D. "What do people know about global climate change?: 1. Mental models." Risk Analysis, 14(6), 959-970, 1994.

[72] Read, D., Bostrom, A., Morgan, M.G., Fischhoff, B. and Smuts, T. "What do people know about global climate change?: 2. Survey studies of educated laypeople." Risk Analysis, 14(6), 971-982, 1994.

[73] Reynolds T.W., Bostrom, A., Read, D. and Morgan, M.G. "Now What Do People Know About Global Climate Change? Survey Studies of Educated Laypeople." Risk Analysis, 30(10), 1520-1538, 2010.

[74] Trenberth, K. E., Fasullo, J. T., & Shepherd, T. G. (2015). Attribution of climate extreme events. *Nature Climate Change*, *5*(8), 725.

recent years, increasing their severity and frequency.[75,76] Yet as suggested by the flooding in New York, the weather extremities climate change is bringing are still likely to exceed many plans and expectations.

Failures to Forestall Catastrophe: Social and Behavioral Sciences Insights and Opportunities

Despite the phenomenally improved forecasts that government research investments have enabled over recent decades, we have failed to forestall catastrophic damages to many of our communities from hurricanes, floods, heatwaves, droughts, and wildfires. There is no single reason for this:

- Interdisciplinary policy research by economists and atmospheric scientists has estimated that hurricanes striking the U.S. cause over ten times the damage that would be caused by an equivalent storm striking another OECD country, due to lack of adaptation in the U.S.[77] In other developed countries, higher income tends to be associated with decreased damages, but not in the U.S.
- In a 2013 survey of U.S. residents in fire-prone areas who had faced fire events (including in Chelan County, WA, Horry County, SC, and Montgomery County, TX) most reported that they had decided to wait and see (65%), rather than evacuate early (24%); 11% had decided to stay and defend.[78] In this study, as in others, receiving an official cue, such as a voluntary or mandatory

---

[75] National Academies of Sciences, Engineering, and Medicine. 2016. *Attribution of Extreme Weather Events in the Context of Climate Change*. Washington, DC: The National Academies Press. doi: 10.17226/21852.

[76] According to Munich Re, the number of extreme weather events globally has more than tripled since the 1980s (https://natcatservice.munichre.com/events/1).

[77] Bakkensen, L.A. and Mendelsohn, R.O., 2016. Risk and Adaptation: Evidence from Global Hurricane Damages and Fatalities. Journal of the Association of Environmental and Resource Economists 3(3): 555-587.

[78] McCaffrey, S., Wilson, R., & Konar, A. (2018). Should I stay or should I go now? Or should I wait and see? Influences on wildfire evacuation decisions. Risk analysis, 38(7), 1390-1404.

government evacuation notification, increased the odds of having evacuated. In the Camp fire, many Paradise residents never received an official evacuation order; 84 people died.

- In superstorm Sandy, despite accurate forecasts, 72% of the 240,000 residents living in mandatory evacuation zones in New Jersey decided to stay in their homes. Across the northeastern U.S. coastline, 117 people died in Sandy, with the most common cause of death being drowning. Half of these drownings were in flooded homes where mandatory evacuation orders were in place a day before the storm's landfall. Red Cross volunteers also noted about the superstorm Sandy drownings that people were "afraid of looters," "thought Hurricane Irene was mild," and "unable to leave because did not have transportation."[79]
- Despite the demonstrably increased influence of online and social media in this era,[80,81] police and fire were largely absent from online media during superstorm Sandy, and government largely absent from Twitter.[13]
- More than five million car crashes happen each year in the U.S., and weather contributes to over a fifth of them. Almost six thousand people are killed annually due to weather-related vehicle accidents, which is almost ten times the 600 adverse weather fatalities that are not related to vehicle accidents. While impressive advances have been made in traveler information systems, these vary widely from state to state, suggesting the need for research on how best to convey the impact and risks of extreme weather events in these systems. Further, while there have been advances in

---

[79] CDC. (2013). Deaths associated with Hurricane Sandy—October-November 2012. Morbidity and Mortality Weekly Report, 62(20), 393. See the discussion in NASEM.

[80] Demuth, J. L., Morss, R. E., Palen, L., Anderson, K. M., Anderson, J., Kogan, M., ... & Henderson, J. (2018). "Sometimes da#beachlife ain't always da wave": Understanding People's Evolving Hurricane Risk Communication, Risk Assessments, and Responses Using Twitter Narratives. Weather, climate, and society, 10(3), 537-560.

[81] Lachlan, K. A., Spence, P. R., Lin, X., & Del Greco, M. (2014). Screaming into the wind: Examining the volume and content of tweets associated with Hurricane Sandy. Communication Studies, 65(5), 500-518.

crowd-sourced data and use of citizen reporting about road conditions, little is known about how drivers value and use them.[82]

- Decades of warning studies show that when a storm, heatwave or wildfire threatens people need to know what to do to be safe, how to do it, and the time until impact of the event. Should they evacuate? If so, how, and when? Where is it safe for them to drive? In addition to key content, the social and environmental contexts affect warning effectiveness, as do the source of the warning, characteristics of the person receiving the warning, and the message delivery method.[83] Although there is a substantial body of knowledge about what makes warnings effective, new challenges have emerged with Wireless Emergency Alerts (WEA), social media, and the entire rapidly evolving ecosystem of information and communication technologies and practices.
- People intuitively understand that there are uncertainties in weather forecasting. Further, the careful experimental research that has been conducted to date shows that people can make better decisions if they are given explicit uncertainty information based on the best scientific forecasts, and tailored to their decision context. They do not, however, always interpret visual and other forecast uncertainty information in the way that forecasters and emergency managers wish or expect.[84] Numerous studies and editorials have highlighted misinterpretations of the cone of uncertainty that the National Hurricane Center uses to show the probable track of the center of a tropical cyclone. For example,

---

[82] Chapter 4 in National Academies of Sciences, Engineering, and Medicine. 2018. Integrating Social and Behavioral Sciences within the Weather Enterprise. Washington, DC: The National Academies Press.

[83] National Academies of Sciences, Engineering, and Medicine. 2018. Emergency Alert and Warning Systems: Current Knowledge and Future Research Directions. Washington, DC: The National Academies Press. https://doi.org/10.17226/24935.

[84] Savelli S and Joslyn S. 2013. The advantages of predictive interval forecasts for non-expert users and the impact of visualizations. Appl Cognitive Psych 27:527–41. See the discussion in: Bostrom, A., Joslyn, S., Pavia, R. Hayward Walker, A., Starbird, K., & Leschine, T.M. (2015) Methods for Communicating the Complexity and Uncertainty of Oil Spill Response Actions and Tradeoffs, Human and Ecological Risk Assessment: An International Journal, 21:3, 631-645, DOI: 10.1080/10807039.2014.947867.

sometimes it is interpreted as the area likely to be affected. But less has been said about how specific misinterpretations influence decisions, which is as or more important to study and understand. There have been suggestions that it might be better to present a spaghetti-type diagram of multiple possible paths for the hurricane center. Presenting an ensemble of paths, that is, a suite of possible future paths, may reduce the tendency for people to see the cone itself as a spatial representation, for example of the size or intensity of the storm. But related research suggests an ensemble graphic could pose other challenges, such as people putting too much weight on a specific possible path.[85]

These examples illustrate a few of the ways that social and behavioral sciences can help identify where there are opportunities to save lives and property, and how to best realize those opportunities. To protect lives and property, and to realize the full value of the investments we've made in the physical sciences, we need to invest in the social and behavioral sciences of extreme weather and climate change.

Beyond Messaging Individuals: Integrating Social and Behavioral Sciences throughout the Weather Enterprise

This research is needed to inform not only personal decisions and behaviors, like evacuation, but also the decisions of emergency management and planning organizations, and of the professionals who

---

[85] K. Broad, A. Leiserowitz, J. Weinkle, M. Steketee, Misinterpretations of the Cone of Uncertainty in Florida During the 2004 Hurricane Season, 88 Bulletin of the American Meteorological Society, 2007, pp. 651–667.
Ian T. Ruginski, Alexander P. Boone, Lace M. Padilla, Le Liu, Nahal Heydari, Heidi S. Kramer, Mary Hegarty, William B. Thompson, Donald H. House & Sarah H. Creem-Regehr (2016) Non-expert interpretations of hurricane forecast uncertainty visualizations, Spatial Cognition & Computation, 16:2, 154-172, doi:10.1080/13875868.2015.1137577.
Wu, H. C., Lindell, M. K., & Prater, C. S. (2015). Strike probability judgments and protective action recommendations in a dynamic hurricane tracking task. Natural Hazards, 79(1), 355-380.

develop and operate our forecast and warning systems, and the professionals and volunteers who oversee and assist with response and recovery. Such research opportunities are illustrated in Figure 1, which is suggestive rather than comprehensive.

People tend to be more prepared for an event when they have prior experience of it, but the type of experience they've had makes a difference.[86] While a plurality of people in the U.S. have long thought climate change will lead to more extreme weather events, their experiences may not be predictive of the weather extremities climate changes will bring. In my home state of Washington, the Climate Impacts Group has highlighted 2015 as a year that may presage climate change. In 2015 snowpack was 70% below normal, leaving the state with irrigation shortages, agricultural losses, fish die-offs, and problems stemming from the state's reliance on hydropower. More shocking for residents was the wildfire, as 2015 was the most severe wildfire season on record for the state.[87] In a more recent example of unexpected extremes, in Texas, tropical depression Imelda dumped three feet of rain in 24 hours, which caught people by surprise, despite Harvey.

---

[86] E.g., Lazo, J. K., Bostrom, A., Morss, R. E., Demuth, J. L. and Lazrus, H. (2015). Factors Affecting Hurricane Evacuation Intentions. Risk Analysis, 35:10, 1837–1857. doi: 10.1111/risa.12407.

[87] Engel, R. A., Marlier, M. E., & Lettenmaier, D. P. (2019). On the causes of the summer 2015 Eastern Washington wildfires. *Environmental Research Communications*, *1*(1), 011009.

Snover, A.K., C.L. Raymond, H.A. Roop, H. Morgan, 2019. "No Time to Waste. The Intergovernmental Panel on Climate Change's Special Report on Global Warming of 1.5°C and Implications for Washington State." Briefing paper prepared by the Climate Impacts Group, University of Washington, Seattle. Updated 02/2019.

May C., C. Luce, J. Casola, M. Chang, J. Cuhaciyan, M. Dalton, S. Lowe, G. Morishima, P. Mote, A. Petersen, G. Roesch-McNally, and E. York, 2018: Northwest. In Impacts, Risks, and Adaptation in the United States: Fourth National Climate Assessment, Volume II [Reidmiller, D.R., C.W. Avery, D.R. Easterling, K.E. Kunkel, K.L.M. Lewis, T.K. Maycock, and B.C. Stewart (eds.)]. U.S. Global Change Research Program, Washington, DC, USA, pp. 1036–1100. doi: 10.7930/NCA4.2018.CH24.

Figure 7. From the National Academies of Science, Engineering and Medicine report, Integrating social and Behavioral Sciences in the Weather Enterprise, Figure 2.1.[88] Stages of communication and decision support that must be addressed under the Weather Ready Nation paradigm, with examples of how social and behavioral sciences (SBS) research can provide critical insights and understanding in each of these stages.

Much remains to learn about how best to communicate forecasts and forecast uncertainties in these circumstances. There is a very large need for additional empirical research on communicating uncertainty for different decision contexts, research that brings social, behavioral, and other scientists together to determine how climate and weather information can

[88] National Academies of Sciences, Engineering, and Medicine. 2018. *Integrating Social and Behavioral Sciences within the Weather Enterprise*. Washington, DC: The National Academies Press.

most effectively be integrated, analyzed and delivered to help forecasters, emergency managers, planners, drivers—indeed, all of us—make better decisions.

Building Leadership and Capacity

The *Weather Research and Forecasting Innovation Act of 2017* (Pub. L. 115-25) as amended by the *National Integrated Drought Information System Reauthorization Act of 2018* (Pub. L. 115-423) is a big step in the right direction, including, for example, specific attention to "Improving the understanding of how the public receives, interprets, and responds to warnings and forecasts of high impact weather events that endanger life and property" [Section 102(b)(2)]; incorporating risk communication research to create more effective hurricane watch and warning products [Section 104(b)(3)]; and mandating collaboration between the public and private sectors to identify the research necessary to enhance the integration of social science knowledge into weather forecast and warning processes, in Section 105(4)).

The National Science Foundation, the National Oceanic and Atmospheric Administration, and other agencies appear to have increased their investments in social and behavioral sciences addressing weather hazards over the last decade, but these investments have been highly variable, and have constituted only a very small proportion – on the order of less than 10% – of their weather-related research investments to date. Such investments have funded pilot programs like the Collaborative Adaptive Sensing of the Atmosphere (CASA) Dallas-Fort Worth Living Lab program, which provides timely, tailored, human-scale forecasts on personal devices, and collects input from users to achieve continuous improvement. While recent initiatives such as *Convergence,* and *Coastlines and People (CoPe),* at the National Science Foundation appear promising, other more obviously closely related funding initiatives, such as Hazards Science, Engineering and Education for Sustainability (SEES), have ended. The investments to date have not comprised the sustained resources required to achieve the advances our nation needs from the social and behavioral sciences in the interdisciplinary domain of weather and climate hazards and risks, or to encourage newly trained scholars to commit to work in it.

In order to fully realize these and other life-saving advances nationally and to achieve international leadership on understanding, forecasting and communicating extreme weather and climate will require scientific leadership and capacity building in the social and behavioral sciences across the public and private sectors, as well. Congressional support is essential. Successfully integrating social and behavioral sciences into an agency or other organization requires senior-level agency leadership,[89] high-level staff to coordinate top-down, bottom-up staff-led working groups, and a commitment to building and sustaining social and behavioral science capacity, through both professional development as well as education, training, and hiring. Detailed suggestions on each of these points can be found in the National Academies 2018 report on Integrating Social and Behavioral Sciences in the Weather Enterprise.

Ann Bostrom is the Weyerhaeuser endowed Professor of Environmental Policy at the Daniel J. Evans School of Public Policy & Governance, University of Washington. She studies risk perception, communication, and decision-making under uncertainty, currently in the contexts of climate change, hurricanes, earthquakes and tsunamis. Bostrom is an elected fellow and elected member of the Board of Directors of the American Association for the Advancement of Science, an elected fellow and past president of the Society for Risk Analysis, from which she received its Chauncey Starr Distinguished Young Risk Analyst Award for her work on mental models of hazardous processes. She is also an elected member of the Washington State Academy of Sciences, an associate editor for the *Journal of Risk Research*, and on the editorial boards for *Risk Analysis* and *Environmental Hazards*. Bostrom co-directed the Decision Risk and Management Science Program at the National Science Foundation from 1999-2001. She currently serves on Advisory

[89] Fischhoff, B. (2017). Breaking ground for psychological science: The US Food and Drug Administration. American Psychologist, 72(2), 118.

Committees for the Social Behavioral and Economic Sciences and for Environmental Research and Education at the National Science Foundation, on the National Oceanic and Atmospheric Administration Science Advisory Board Environmental Information Services Working Group, and on the Advisory Committee for the National Center for Atmospheric Research Mesoscale & Microscale Meteorology Lab. At the University of Washington she serves on steering committees for EarthLab and the Center for Studies in Demography and Ecology, and is on the governing board of the Program on Climate Change. Bostrom co-chaired (with Bill Hooke) the National Academies of Science, Engineering and Medicine *Committee on Advancing Social and Behavioral Science Research and Application within the Weather Enterprise* (consensus report published 2018). She completed postdoctoral studies in Engineering and Public Policy at Carnegie Mellon University and in cognitive aspects of survey methodology at the Bureau of Labor Statistics, and holds a PhD in public policy analysis from Carnegie Mellon University, an MBA from Western Washington University, and a BA from the University of Washington.

* * *

Chairwoman JOHNSON. Thank you very much.

We will now begin our first round of questions, and I will recognize myself for 5 minutes.

As the scientific consensus continues to show certain extreme weather events are directly impacted by climate change, but there remains uncertainties in the attribution of climate change and to other extreme events. So my question to all of you, what are the most pressing research needs to better understand the physical processes that drive extreme weather events?

Dr. SHEPHERD. Well, I'll start the question, and I'll focus on hurricanes because what we know about hurricanes from an extreme-event standpoint is that our forecast tracks have improved remarkably in the last several decades. Where we still lag is in intensity forecasts. And the reason that is basic meteorology, the intensity processes associated with hurricanes involve things happening within the inner core of the hurricane and the eye wall of the storm beneath the storm, and so we need observations in those regions.

Tracking or the prediction of track of a hurricane is governed by larger-scale processes. Where is the Bermuda High sitting over the Atlantic? Will that storm be pushed into Florida or North Carolina by that steering current?

But that energy released—I often describe hurricanes like a car engine. There's a lot of energy being released in the eye wall as those thunderstorms are rising. And we're not observing those processes. Our models aren't scaled to represent those processes.

I mentioned in my opening remarks the University of Georgia is testing, through some NOAA funding, these robotic underwater gliders, and they are actually measuring warm pools, deep ocean water. That type of information can be very useful in understanding the intensity processes.

So to your question, Chairwoman, I think in terms of hurricanes, the intensity problem is still a very large challenge for us. I would also say on the rainfall aspect of the hurricanes we're seeing more rainfall in Hurricanes Harvey, Florence, Imelda. We warn on wind with our Saffir-Simpson scale. That's a wind scale.

And so I think in terms of the communication challenges that we heard earlier, and it echoes with what my colleagues said, we need to think about ways of communicating to the public that rainfall and water is what kills most in storms. And so I think there's a combination of intensity and the communication of risk as we see this new normal in extreme events because past experience with a hurricane doesn't sort of predict your future outcome from a new storm because this is a new normal.

Chairwoman JOHNSON. Thank you.

Dr. DONE. So scientists are on the cusp of something of a revolution in how we are able to simulate extreme weather events, so in the next 2 years or so we expect we'll be able to see these devastating events in very fine levels of granularity over decades. So this really represents a transformation in our ability to understand the processes and the interactions between the weather and the climate. So this represents a new horizon of understanding, which should feed into improved forecasts on the seasonal to decadal scales.

Dr. MOORE. I will buildupon what Marshall Shepherd mentioned and focused on hurricanes. I will focus on tornadoes. And I'd like to just observe that we all know the phrase Tornado Alley. That phrase is being replaced these days by Dixie Alley. There appears to be a migration of very large tornadic activity to the southeast. These tornadoes tend to be larger, and they tend to occur at night.

Why is this happening? Well, we don't know. But we know it is happening. The observational infrastructure of the southeast presents a more difficult situation than in Oklahoma because of the terrain. And yet the observational situation in the southeast is less dense, so you have a terrain problem with a less-than-dense observational network, and that's going to need to be addressed. In particular, I think a national mesonet—there really needs to be a national mesonet at the standards of Oklahoma and in fact maybe have to exceed those in these more terrain-challenging areas.

Second, we're going to have to have a radar network that is equal to the task of the 2040s and 2030s, which is, as I mentioned, going to require work now.

Finally, I think in the models, not only do the models have to be better in their physical properties, as Marshall mentioned, I think they've got to also buildupon a better observation basis of the boundary layer, the lower part of the atmosphere.

Chairwoman JOHNSON. Thank you. My time is expired. Thank you. Mr. Lucas.

Mr. LUCAS. Thank you, Madam Chairman.

Dr. Moore, let's expand for a moment on your comments in your opening date testimony and just now about the mesonet system. And for our friends in the room who have never been exposed to it, this is a network of weather data gathering sites all over the great State of Oklahoma. There's, what, probably 30, 40, 50 information points from subsurface soil temperature to wind, to a variety of things. And it is fully accessible to everyone, correct?

Dr. MOORE. Yes, and it's well over 100.

Mr. LUCAS. Well over 100 sites. So can you discuss for a moment teeing off of what you've said about the challenges in other parts of the country about how the model could serve and could be a model for the rest of the country in enhancing our forecasting?

Dr. MOORE. One of the areas that the mesonet—first of all, let me back up from that and say when we look at satellites, particularly those satellites that are in polar orbit from pole to pole, that's really collecting global information that is extremely important for numerical weather prediction, 3, 4, 5 days out. But when you're into the endgame of a very intense situation, that's when the geostationary satellites and the mesonet really begin to play their role because they're giving you timely, in situ information in the battle zone in where the weather event is happening. And that's proven extremely valuable in Oklahoma.

I should also mention there are many other benefits. We have something called OK-FIRE, which basically allows a farmer to come on to the mesonet and query with their ZIP Code, can I burn the field today? Is it OK? And so the soil temperature information, the wind information is allowing the people to make better decisions. Now, these are weather-related but is an additional benefit to the predictive capability, particularly in that endgame situation.

Mr. LUCAS. And that information is available to everyone.

Dr. MOORE. Yes, it's on the web and, I mean, this is something that when I moved to Oklahoma I really didn't know about this, and all of a sudden I thought, well, this is really important. It allows me to have in situ information all over the State. And this is something that we really need at the national scale. And as I also mentioned, we need to take that information into the lower part of the boundary, into the lower part of the boundary layer of the atmosphere to get that in situ information that is really needed for weather prediction as you get into an endgame situation.

Mr. LUCAS. To continue, Dr. Moore, I'm of course very proud to have introduced the *Weather Act of 2017*, which was signed into law a little more than a couple of years ago to provide weather research and forecasting improvements, weather satellite data innovation to try and drive things forward in a coordinated fashion. In your experience have the goals of the *Weather Act* been met? Tell us about the implementation.

Dr. MOORE. No, I don't think they have. I think that this is going to be challenging. We have operated in a certain way in terms of improving our models for a long time, and we are going to need to think afresh about how we do this.

As I mentioned, I believe we have the greatest scientific talent in the world in the United States in the area of meteorology, but it needs to be marshaled and applied to this national and global challenge. There are really many opportunities out there to allow this to happen, and I think the EPIC initiative is really at the fundamental level. But there are other elements that are right teed up for this, for instance, cloud computing. This may be a whole new way in which teams can form around critical areas and involve the cloud in a collaboration in terms of improving the models because we've got to make that improvement.

Mr. LUCAS. Yes. Speaking of that, and I address this to the whole panel, and we've all referenced EPIC several times. Would anyone care to expand on how EPIC will help improve that weather forecast, that scale when it comes to extreme weather?

Dr. MOORE. I'll let my colleagues speak.

Mr. LUCAS. Absolutely.

Dr. DONE. Yes, I think one of the values of EPIC is the two-way interaction between the research and the operations, so it's critical for scientists to understand the forecast needs and to enable the science to be relevant, so this continual iteration, this two-way flow can really accelerate and catalyze the research to operations.

Dr. SHEPHERD. And I would just add to that, just like any good decisionmaking and problem-solving session, when you have more voices and ideas at the table, you have more diversity of thought to challenges. And so I think by bringing the academic community and the best ideas from academia and other partners into the process of the American modeling system, you're just bringing more good ideas to the table.

Mr. LUCAS. Well put, Dr. Shepherd. My time is expired, Madam Chair.

Chairwoman JOHNSON. Thank you very much. Mr. McNerney.

Mr. MCNERNEY. I want to thank the Chair. I thank the witnesses for your testimony, very interesting and important.

Dr. Done, could you explain the difference between climate change and climate variability and how we see the climate change signal in our rising losses apart from climate variability?

Dr. DONE. Yes. So we know that extreme events have varied on different time scales throughout history. So, for example, people have taken sediment cores along our coasts and along global coasts to look at the signatures in hurricanes, and they have varied on time scales of decades and hundreds of years, for example.

So what's different today? Well, we know that we are unable to replicate observed trends in extreme events without invoking human influence in our model simulations. So we rerun the history with and without human activities in our models or the impact of human activities, and there's a clear signal that there's a human influence.

And let me give you an analogy of climate variability in climate change. So if you imagine a dog walker taking a dog for a walk across a park, the line traced out by the dog can be thought of as climate variability. But fundamentally, the line trajectory by the dog walker is climate change. So climate variability fundamentally has to follow climate change over a long enough distance or time period.

Mr. MCNERNEY. Well, would you elaborate a bit on how risk management can be integrated with climate modeling?

Dr. DONE. Yes. Yes, in my experience as a Willis Research Fellow has shown me the value of, again, this two-way understanding. So I talk to risk managers to understand their key needs. For example, one of them, say, in California is the need to understand the successive atmospheric rivers or these winter storms in California. So I can go back to the data and pursue science discovery around that topic to try and improve our understanding of these sequences of events that will lead to better forecasts and better understanding of the overall risk.

Mr. MCNERNEY. Well, thank you. Dr. Shepherd, how does NCAR and NOAA change forecasting for extreme events such as 100-year flooding since we're entering a new era with changing climate?

Dr. SHEPHERD. I think that's a very important question because that's one of the messages that I often convey. Your point about a new era is very important. And what we saw with NOAA quietly I might add, a lot of people aren't familiar with the fact that they are updating the flood frequency and rainfall maps because we are seeing changes.

I think what we need going forward in addition to systems like the EPIC, which you hear us all mentioning because it is so important, we need to fundamentally understand sort of how our models, how our observations are framed to understand these extreme and compound extreme events. So, for example, with Imelda just recently in Texas it was a tropical depression when it dumped 3 feet of rainfall, but that wasn't the full story. It wasn't just the tropical depression. There was also something in meteorology we call a trough that which is part of our large-scale mid-latitude synoptic system that came into the State of Texas as well and sort of amplified the situation. And so although we saw 1 to 2 feet of rainfall potential out of Imelda, it dumped 3, in some cases 4 feet of rain, close to 4 feet of rain.

So as NCAR, NOAA, and other colleagues start to develop models, there will need to be assessment of how we're representing the extreme end or the tail end of these extreme events. And I think that's an area of needed research.

Mr. MCNERNEY. Excellent. Would you comment on the accuracy of climate modeling?

Dr. SHEPHERD. I'm glad you asked that question because I come across people that often—you know, they're sort of stuck in this mode that the climate models aren't accurate or they can't tell us anything about climate. And one of the big challenges I see is that people confuse weather and climate models. They're apples and oranges. A weather model is trying to depict the state of the atmosphere 1, 2, 7, 10, 14 days out. A climate model is not trying to tell us what the atmosphere looks like on

Wednesday at 2 p.m. in 2075. It's trying to predict the state of the climate system.

And they do quite a good job because we now have a modeling system that represents the full Earth system. It represents the cryosphere, the atmosphere, the ocean, the landmass. We have evolved from a generation of models in the 1970s and 1980s where there might have been some concern about the accuracy.

Are there still uncertainties in the climate models? Absolutely. There is still some uncertainty about the atmospheric aerosols, the pollution and particles in the atmosphere. There's still some uncertainty about the representation of clouds in those models, but there is certainly enough information to make a decision.

Uncertainty gets a bad rap. Uncertainty is not we don't know. It just means uncertainty. If I tell you there's an 80 percent chance of rain tomorrow, there's uncertainty in that information, but there's certainly enough information for you to grab an umbrella, too.

Mr. MCNERNEY. So in Dr. Done's analogy, you can kind of tell where the dog walker is going to go——

Dr. SHEPHERD. I——

Mr. MCNERNEY [continuing]. Even though the wiggles are going to be——

Dr. SHEPHERD. In terms of the climate dog, I know where that dog's headed.

Chairwoman JOHNSON. Thank you very much. Mr. Weber.

Mr. WEBER. Thank you, Madam Chair. And I do have a report submitted from CenterPoint Energy I'd like to get into the record.

Chairwoman JOHNSON. Without objection.

Mr. WEBER. Thank you. I'm fascinated by a couple things. The mesonet system I was unaware of. It looks like there's 110 sites. Is this cellular-based? Do we know? Dr. Moore, I guess that would be for you. You seem to know more. You see what I did there?

Dr. MOORE. Yes, set me up. Yes, it is, and we actually work with the Oklahoma State Patrol and their communications system to make sure that we have very fast information. Every 5 minutes it's updated, and that information then is available to anyone who logs on to the internet system.

Mr. WEBER. And there is an App for it?

Dr. MOORE. Yes.

Mr. WEBER. Do you have that App on your phone?

Dr. MOORE. Oh, yes.

Mr. WEBER. OK. All right. Dr. MOORE. Everyone does.

Mr. WEBER. Everyone? I don't have it on my phone. Dr. MOORE. Come to Oklahoma.

Mr. WEBER. Come to Oklahoma. That's north Texas we call that. And, Dr. Bostrom, you talked about the need to invest in social and behavioral sciences. What do you mean by that?

Dr. BOSTROM. There's a broad range of sciences that can help us understand both management opportunities and how people undertake management opportunities. For example, if we think about EPIC, there's a lot of interest in community modeling and how EPIC can advance what we know about modeling in the atmosphere.

But it's very obvious from the discussion around EPIC that there are issues about how the community modeling itself should happen, what should be the procedures that guide it, and that kind of thing.

There's a lot of research that can help inform what kinds of processes or methods for coordinating this could work better.

There's also, of course, the obvious roles of behavioral and social sciences in improving communications and improving our understanding of how people assess their risks and act on them.

Mr. WEBER. That's a pretty tall order because I've lived on the Texas Gulf Coast in a 20-mile radius 66 years, 8 months, and 26 days. Who's counting? And I've been through a whole lot of hurricanes. And you see from Hurricane Ike, which most of you all would recognize, and then also obviously Hurricane Harvey and Tropical Storm Imelda was ground zero in mine and Dr. Babin's district here in the last 2 years, and people don't take that serious when they say it's time to evacuate.

If there's going to be some kind of certainty that we can come out with, how do you convince people—a lot of people evacuated, for example, during Rita and Katrina and some of the others, and the Houston highways were just absolutely inundated—pardon that use of the expression. And people couldn't get out. And there were several deaths on the highway and they ran out of fuel and all kinds of stuff. And then a lot of them felt like they left needlessly.

So how do you convince people through social and behavioral sciences that they can now trust our forecasts? How do you get to that point?

Dr. BOSTROM. There's evidence that people pay a lot of attention to warnings and official warnings in particular and that they're very important in influencing people's actions. People's actions are constrained not only by what they hear from the official warnings but also by their own response opportunities, what they can afford——

Mr. WEBER. Their own experiences?

Dr. BOSTROM. Their own experiences as well and what their neighbors are doing. There's all kinds of influences. So just being able to persuade people that they should evacuate is not the only thing that's going on.

Mr. WEBER. You mentioned that you study risk perception on page 1 and decisionmaking.

Dr. BOSTROM. Yes.

Mr. WEBER. Who is your target audience when you study that?

Dr. BOSTROM. It depends on the topic and the funding and——

Mr. WEBER. OK. Well, let's talk about storm decisionmaking and evacuation——

Dr. BOSTROM. Yes.

Mr. WEBER [continuing]. And how much funding?

Dr. BOSTROM. So storm decisionmaking and evacuation, the only work that I've done on that has been in the context of flash flood and hurricane forecast and warning systems. And there we were looking—it was a National Science Foundation-funded project, so our target was understanding what it was about the forecast and warning system that might be improved and how it might be improved in terms of providing accessible and available information to the different parties in the system to be able to make good decisions. So that meant looking both at emergency managers and what they need. It also is how broadcasters get information on how they use it to talk to people.

Mr. WEBER. Sure. OK. Well, I need to move on here just one more question for Dr. Shepherd down here. When you're talking about predicting weather patterns and currents and inside of storms you talked about you couldn't get to the eye wall there, couldn't into those things, how do you with any degree of reliability predict wind currents and what the model airplane people call thermals, for example? Stuff just came up from the Earth. How do you predict those?

Dr. SHEPHERD. Well, you know, the atmosphere is a dynamic and thermodynamic system with moist processes in it, so our models actually can identify thermals in those processes, but what I was talking about, those thermals if you will in the eye wall of a hurricane, we can get information from the brave hurricane hunters that fly into those storms. There are satellite systems. I actually was the deputy project scientist for one when I spent 12 years at NASA called the Global Precipitation Measurement mission or GPM. It actually has a radar in space, and so we can see something called hot towers, these bubbling thermals that you mentioned. And they provide clues that a lot of energy is being released in that hurricane and we might see a strengthening of the storm.

Here's the problem. That instrument is only on a polar-orbiting satellite that only gets a snapshot of the storm every now and then. So we need a generation of technology that's giving us more robust spatial coverage, perhaps a geosynchronous radar or a network of cube or small sats——

Mr. WEBER. OK. Very quickly. I'm about out of time. But when you say it only gets a snapshot, those orbit the Earth every 3 hours, 6 hours?

Dr. SHEPHERD. So polar-orbiting satellites are in low-Earth orbit whereas our geosynchronous satellites are at about 35,000 kilometers up, so they're just staring at the same spot on the Earth the entire time. So a couple of times a day perhaps or—it just depends on the type of orbit the satellite is in. So we don't always have that information. We need more sustained information like that, and that's really the value of some of the advanced observation systems that NASA is looking at in its decadal survey.

Mr. WEBER. Thank you. Madam Chair, thank you. I yield back.

Chairwoman JOHNSON. Thank you very much. Ms. Bonamici.

Ms. BONAMICI. Thank you so much. Thank you to all of our witnesses here today. I was pleased to work with Ranking Member Lucas on the *Weather Research and Forecasting Innovation Act of 2017*, which, thank you for the acknowledgment, many of you took important steps to strengthen the capabilities and communication of weather forecasts. We know extreme weather events like Hurricane Dorian, for example, remind us of how important that legislation was but also the value of the National Weather Service accurately and effectively communicating information.

And if erroneous information is reported, it's the responsibility of the National Weather Service to refute it. It's exactly what the Birmingham office did earlier this month. And I know it's not the focus of today's hearing, but it's really important to acknowledge how those public servants help defend scientific integrity and what unfortunately became a political moment.

I want to follow up on Mr. Weber's question to Dr. Bostrom. During the legislative hearings on the *Weather Research and Forecasting Innovation Act*, we had extensive conversations about how forecasts will not adequately serve the needs of the public unless there are effective communication systems. The bill directed NOAA to do more research, to listen to experts, to improve its risk communications techniques. It is my understanding that progress has been made, but the pace has been slow.

In your testimony I thought it was really compelling where you talked about with Sandy, 72 percent of the residents were in mandatory evacuation zones in New Jersey stayed in their homes. And sadly, 117 people died. So this is an issue that affects human life, as well as property. There's always some uncertainty in forecasts, we know that. So what research gaps still persist in our understanding of how forecasters can effectively communicate in light of the uncertainty? And how can we better integrate social and behavioral sciences in the conversation about extreme weather events? Dr. Bostrom?

Dr. BOSTROM. Thank you for this question and for acknowledging the importance of the *Weather Research and Forecasting Innovation Act.* It's really an important step forward. But, as you said, progress has been slow. And while we have made progress in the social and behavioral sciences in this context over the last couple of decades, it has been very variable and limited funding devoted to this. We need research both on how the weather enterprise as a whole works.

So, for example, if we think about what happened with Sandy, there are recent studies that have shown that people were paying a lot of attention—especially people in evacuation zones were paying a lot of attention to what was going on in various social media. They get information from a lot of different sources, and they're also paying attention to the environmental cues that are going on around them.

Further, as mentioned previously, they're paying attention to their previous experiences. So if they don't——

Ms. BONAMICI. Right.

Dr. BOSTROM [continuing]. Expect a storm as bad as what they've said and as I believe it was Dr. Shepherd said, the storm surge is often the most dangerous. And we know from prior experience that—in the research field as well that people don't anticipate the dangers of storm surge still. The storm surge products that have come out communicating storm surge are relatively recent, and people didn't anticipate that things would be as bad as they were.

So we need to understand both how to use those communication systems better, how forecasters can work with these teams better to understand what emergency managers need, and we also need to understand better ways of communicating to people the dangers of storm surge and what those can bring.

Ms. BONAMICI. Thank you. Still some work to do.

Dr. Shepherd, yesterday, the Intergovernmental Panel on Climate Change released a special report on the ocean and cryosphere in a changing climate. The report found that warming oceans and projected sea-level rise will result in increasingly severe extreme weather events. And in your testimony you discuss how disparities in income, social status, and other factors result in hurricane, flood, and heatwaves having disproportionate and adverse effects on certain marginalized populations. We also know that extreme weather events differ based on geographical location. Mr. Lucas is in Oklahoma, and I'm in the Pacific Northwest. In the Gulf it might mean more hurricanes. In northwest Oregon it could be more intense rain, heatwaves, severe drought, wildfires, reduced snowpack. So how can we more effectively assist our local communities in assessing the scientific information they need to make informed decisions for resiliency planning, which is so critical?

Dr. SHEPHERD. Thank you for that question. I live in the south where we get every single type of extreme event. It's actually unique in that regard. And we literally get every type. Coupled with that, if you look at the population based in the south, it tends to—particularly some of the southern States—to have low socio-economic status in terms of the wealth gap that I've often talked about in some of my studies, and so that increases vulnerability when we have a heatwave or when we have a Hurricane Harvey. All you have to do is look at the faces staring at us in the Superdome during Hurricane Katrina.

One of the things that we're doing in the State of Georgia is we've stood up, through funding from the Ray C. Anderson Foundation, a stalwart of the business community, is something called the Georgia Climate Project. And that was called out in the National Climate Assessment as a potential best practice, a solutions-oriented effort to connect climate processes at local levels through businesses, through stakeholders, nongovernmental organizations, and just regular people in their communities because there is a disconnect between all of this science jargon and mumbo-jumbo that we talk about as scientists and my aunt who lives in Canton, Georgia, or great aunt who knows none of this terminology but knows that they are experiencing events that affect the cost of cereal they buy or the price of gas when there's a hurricane plowing through the Gulf of Mexico.

So we need to think about and something I want on the record, we need to think about these communities in our country that are extremely vulnerable and at higher risk for these events. And I echo your thoughts on the National Weather Service, and I want to thank, as the former President of the American Meteorological Society, all the men and women of the National Weather Service for what they do.

Ms. BONAMICI. Thank you. I see my time is expired. I yield back. Thank you, Madam Chair.

Chairwoman JOHNSON. Thank you very much. Mr. Babin.

Mr. BABIN. Thank you, Madam Chair. I appreciate that. And, Ranking Member, I appreciate you. Thank you to the witnesses for being here today.

I have the distinct privilege of representing southeast Texas, but unfortunately, that means that I'm far too familiar with these terrible storms and floods. And I continue to see these 2,000-year floods every 2 years. And we've had two sweeping through my district in the past 2 years, almost 2 years ago to the day to when Hurricane Harvey swept through southeast Texas leaving a record rainfall in my district. Tropical Storm Imelda last week flooded the same Texans who were still recovering from Harvey. I've gotten calls from constituents who, just weeks ago, were finally able to move back into their homes when Imelda left them once more flooded out.

Dr. Moore, Tropical Storm Imelda was upgraded from a typical rainmaker to a full-blown tropical storm, and many of my constituents were caught by surprise, including myself, as these flood-waters rose. Can you talk about the threat of rapid intensification, which some of you have addressed already this morning and sometimes just hours before landfall for hurricanes and how forecasters take this into account?

Dr. MOORE. Just building upon what Marshall Shepherd said earlier, that we really had a very unique meteorological condition, and yet we should be able to address unique meteorological conditions. And we have seen a pattern of our models not catching the intensification accurately. We've done better on the landfall prediction, but we don't seem to be catching the intensification, which is, after all, what really matters to those in the landfall region. I think that this is, again, where observations, particularly observations that are contemporaneous with the storm, that that is observations that are persistent is an absolute requirement. And I think that if we had a sounder in geostationary orbit, we would have done better in catching the intensification. It really is a product of the boundary layer in large measure.

And I think that the fact that we failed to catch this intensification has had a counterproductive effect upon the body politic. People begin to say, well, it's uncertain or it wasn't predicted, and that leads to inaction. And so the fact that our models have let us down at certain stages should actually cause us to be more vigilant in terms of taking action. But people will say, oh, well, I've been through this before or there were this was uncertainty or the models are saying different things and they use that as kind of a logic for inaction.

Mr. BABIN. OK. Thank you very much. And, Dr. Sobel, I would say it's fairly common to see headlines claim that hurricane-related inland rainfall and flooding has increased due to climate change. Just earlier this month *The New York Times* published an article that stated, "Some attributes of storms, particularly the increasing amount of rainfall associated with many of them, have reached a very strong consensus."

Dr. SOBEL. Yes.

Mr. BABIN. Yet NOAA concluded an anthropogenic influence has not been formally detected for hurricane precipitation. WMO stated, "No observational studies have provided convincing evidence of a detectable anthropogenic influence specifically on hurricane-related precipitation." USNCA said, "There is a lack of supporting detectable anthropogenic contribution and observed tropical cyclone data." And to me this seems like the very opposite of any type of consensus, yet alone a very strong one. In your opinion why is the public discourse, including from scientists, often at odds with thorough assessments like the ones NOAA, WMO, and others have published?

Dr. SOBEL. Thank you for the question. So I think it's important to understand the meaning of the word detect. I think that scientists in my field—and I include many of the great experts at NOAA who write those statements—but all of us use a very conservative standard. So what detection means is, can you say with 95 percent confidence that the changes we've seen could not have occurred without human influence?

That's a very strong standard. And some of the limitations in that are simply the observations. In other words, you have numbers in intensities of hurricanes and rainfall that are fluctuating up-and-down year-to-year naturally. Any trends are slow upon that. The observations themselves have limitations, as Dr. Moore has said, especially with rainfall, which isn't measured that well through all parts of the hurricane.

So it's a question of—what those statements are saying is not that there aren't changes but that you can't say based on the observations with 95 percent confidence that they couldn't have occurred naturally. But I don't think that's the right question to ask.

So we know that the atmosphere has more water vapor. We know hurricanes are very good at squeezing that out. So from all sorts of evidence we have from our physical understanding and our models show pretty substantial increases of rainfall in hurricanes.

The observations are not contradicting that. They're simply so noisy that you can't pull out that signal. So I like my example of people speaking quietly in a crowded room. If it's loud, it doesn't mean—you know, you can't necessarily say with 95 percent confidence what they're saying, but that's not a reason not to take it seriously. So I think that's where the tension is between the very conservative stance used for detection versus our scientific understanding of what's actually happening.

Mr. BABIN. Thank you very much. My time is expired, which is unfortunate because I have other questions.

Chairwoman JOHNSON. Thank you very much. Mr. Casten.

Mr. CASTEN. Thank you, Madam Chair. Thank you so much to our panelists.

My longtime friend Mike McCracken, who is one of the co-authors of the IPCC report that shared the Nobel Prize with Al Gore, has been pointing out for some time in ways that are very troubling that the historic challenge with our climate forecasts is that they are prone to underpredict the severity of what happens because what we fail to always appreciate is all the complexity of the feedback loops, and almost all the feedback loops are positive. He was out here last week saying how 10 years ago we didn't think the Antarctic ice sheet was going to melt, and now we see it melting and we have to start factoring that in.

That leads to the reality that I think you all have well-articulated, that extreme weather events are becoming more and more common, they are becoming more common than we need to, and I certainly thank you, Dr. Shepherd, for articulating the climate forecasting and the weather forecasting are two different things, but they connect in that extreme weather event moment.

And I was rather concerned when the Trump budget forecast suggested we cut 110 employees from the National Weather Forecast. I raised those concerns to Dr. Jacobs when he was here, who assured me that the computers would be accurate. And you can imagine why, given Mr. McCracken's comments, that makes me a little nervous.

Dr. Bostrom, your research on the importance of humans that understand how humans communicate so that what we want to be heard will be accurately heard I think is critical to making sure that we do that because that's a hard thing to do on a computer.

Notwithstanding all of that, we have good people at the National Weather Forecast. We have really, really impressive models. They're really complicated. They're being adjusted in real time. All of Mike's concerns are being integrated into the next model, and yet those models have to compete with a President with a Sharpie. And that's not a joke. We have real people who are listening to that as well and trying to make decisions.

And, you know, there's a reason why knowingly issuing a false weather forecast is a crime punishable by up to 90 days in jail.

Dr. Shepherd, can you please explain to the Committee what the professional obligation is that exists for forecasters to correct the record when misleading weather information is causing unnecessary panic? And how do you balance that given the innate uncertainty to any forecast?

Dr. SHEPHERD. Well, thank you for the question. And I—again, as someone that has sort of represented the broader meteorological community as the President of the American Meteorological Society, I have talked with virtually every corner of this country in terms of the

National Weather Service forecasters. If you look at the mission statement that they operate under, it is to protect lives and properties.

I won't delve into sort of the political investigations or the political thought here, but I will say this very clearly. The obligation to be clear about the current situation with the meteorological forecasts for an extreme event like a Hurricane Harvey or Hurricane Dorian is of utmost importance. And I believe that's what the National Weather Service Birmingham forecasters were doing. And if that same situation arose again, they should do exactly the same thing.

The second thing I believe is that by mixed messages—and we don't just see this in this particular situation. We as the meteorological community have an extreme challenge because there are all kinds of weather opinions and forecasts on social media and other places, and there always is this running joke as meteorologists, many meteorologists in the room, NOAA, that it must be nice to work in a field where we're wrong half the time and still get paid. Well, that's a myth. We are actually right most of the time but, just like in football, if a field-goal kicker misses an important field goal in the Super Bowl, people remember that even if you made all the field goals all year long. That's the bias that we deal with and the myth.

However, when we start questioning the expert forecast from the National Weather Service or the National Hurricane Center, that undermines and in my opinion jeopardizes safety of our public, of our citizens because if someone now starts to say, oh, well, I don't believe the National Hurricane Center forecasts and that hurricane is headed my way, they may make a poor decision in terms of a decision——

Mr. CASTEN. I'm tight on time, but I want to get one last question in to Dr. Bostrom. Can you please share with us what the research shows about how people's trust in forecasts is impacted when they are confronted with false positives, and how does that threaten public safety the next time we have an extreme weather event?

Dr. BOSTROM. I'm hesitant to speak without my data in front of me, but thank you for the question. We know that trust is an essential component of any kind of communications process, especially with regard to risk. And we also know from the context of tornado discussions and warnings in particular, when people get false warnings, it does cause a decrement in their behavior. But not as much as some people might expect.

That said, I don't think that we're talking about the kinds of situations in those research projects that we have here. It is clear that false alarms do cause a decrement to behavior. Having correct alarms is really important, and so the more correct alarms you have, the more you can offset those effects of false alarms. And people are often very concerned about missed alarms, so that's a really important problem. I hope I've answered your question.

Mr. CASTEN. Yes. Thank you. And I yield back. Chairwoman JOHNSON. Thank you very much. Mr. Rooney.

Mr. ROONEY. Thank you, Madam Chairwoman. And with permission, I'd like to introduce three charts to go with my comment if I might?

Chairwoman JOHNSON. Without objection.

Mr. ROONEY. I'd like to ask Dr. Done and Dr. Shepherd a question about the deep ocean heat content and the relationship of CO_2 to that, and whether or not it's leading to stronger storms and what might happen if we can't arrest the rise in sea temperatures.

This is a chart showing the rise in sea temperatures since the early 1980s, particularly the western Atlantic, western Pacific, and Gulf of Mexico. This is a chart showing the dramatic rise in the deep ocean heat content. And then this third one is one showing the same rise and breaking it out both above and below the 2,300-meter demarcation line. Thank you.

Dr. DONE. Yes, great question. So we know from the data that much of the warming is going straight into the ocean by a large amount rather than warming up our atmosphere. And in fact if all the warming stayed in the atmosphere, our average global temperature would be many tens of Fahrenheit warmer than they are today. So there's huge changes to the heat content in the ocean, and this has ramifications for extreme events.

So we talked a lot today about hurricanes already. They feed off this reservoir of energy, this fuel. So even just a 1-degree increase in the ocean heat content, say, in the Gulf of Mexico has vast impacts on the characteristics of extreme events such as tropical cyclones and attendant flooding, for example.

Dr. SHEPHERD. Added to that, I mean, I think your point is valid. One of the things that most people don't realize is most of the quote/unquote global warming is happening in the oceans. That's about where 90-plus percent of the warming is going. We sort of quibble and argue about that small percentage in the atmosphere, but what I always tell my students at the University of Georgia is that heat in the ocean find its way back to the atmosphere through hurricanes, through changes in weather patterns.

Something we haven't talked much about today but just changes in the heating patterns in the Arctic region, Arctic amplification, that communicates to where we live even if we don't live in the Arctic and we live in Dacula, Georgia, for example, the jet stream patterns that respond to those Arctic amplification changes some of the scientific studies suggest.

So your question about, exactly, that you're pointing to in your graphic there, the thing to understand and why we need global observations is that the Earth system is a connected system, so something happening far off in the Arctic or in the Pacific Ocean can affect where we live.

Mr. ROONEY. Thank you. If I might, one more question for Dr. Bostrom and Dr. Sobel. Coral reefs play a very important part of our protection of shorelines, as well as our ecosystems, and science seems to be proving that stronger storms and ocean acidification are related. And so what I'd like to ask about is what can we do to reduce ocean acidification and to protect our coral reefs? And what is the correlation between ocean acidification and coral bleaching?

Dr. SOBEL. I'm not an expert on corals, so I don't want to give too detailed of an answer, but I do understand from my colleagues who study them that they are in big trouble due to both warming and acidification of the oceans. I can't say the relative roles of both of those, but it's a very serious problem.

And I don't know that there's anything that can be done to stop ocean acidification on a practical level other than putting less carbon into the atmosphere. The oceans are taking up a large amount of that, and the ocean is so enormous that, you know, one can imagine geo-engineering schemes to put something in the ocean to mitigate it, but I don't think those are practical. And the same is true for the warming of the ocean. I really think mitigation of greenhouse gas emissions is the only answer there.

Mr. ROONEY. Here's a picture of a bleached coral, dead. Anybody else have any thoughts about it?

Dr. BOSTROM. I would just echo what Dr. Sobel said, that I'm not an expert in any of those things, but it's very clear that the strategies that we have to mitigate what's going on with coral reefs and other potential living systems that help protect the coasts require reducing greenhouse gas emissions.

And geo-engineering strategies have been tested in the past with limited positive results as far as I understand, and we are in desperate need of new geo-engineering strategies and research on them.

Mr. ROONEY. Thank you very much. I yield back.

Chairwoman JOHNSON. Thank you very much. Mrs. Fletcher.

Mrs. FLETCHER. Thank you very much, Chairwoman Johnson and Ranking Member Lucas, for holding this hearing. And thank you to all the witnesses who are here testifying this morning. Your testimony is really interesting. And I'm sorry I only have 5 minutes. But I do want to follow up on a couple of things.

We've heard from my two neighbors, Dr. Babin and Mr. Weber. We all represent southeast Texas, and I represent a portion of Houston, so these issues are near and dear to my heart and to my constituents who are still recovering from Hurricane Harvey, now from Imelda, and from four 500-year floods in 5 years. So we understand the remapping, the rainfall.

And I think that there are a couple of things that I would just love to hear your thoughts on and get you to touch on. In particular I really liked in your prepared testimony and in your remarks this morning, Dr. Shepherd, the conversation or the comments about how we're messaging and how we're talking about what the real impacts are. And, for example, I just got a news report that, in addition to the other things we know about Imelda, there were nearly 100,000 pounds of toxins released in the air from related incidents to the rainstorm.

So when you talked about how extreme weather affects our water supply, public health, infrastructure, energy systems, et cetera, I think it be useful to get your thoughts on what those things are and how we can talk about those, that it isn't—while we all are concerned about some of the long-term effects, there are also immediate impacts that I think are useful to communicate.

Dr. SHEPHERD. Sure, absolutely. I think it's critical. I think particularly if I were sitting in the shoes you all are sitting in because constituents resonate with how much they're paying for cereal or their gas prices or whether their child is vulnerable to a particular disease from a mosquito that couldn't live where they live now but can because of changing climate envelopes.

So I just think—and I've written about this in an article I wrote in *Forbes* magazine about reframing climate not as the sort of esoteric or sort of very nebulous issue but about kitchen-table issues. And there are so many of them. There's a colleague that has—many of us are—not me but many of us fly out almost daily or fly on airplanes. There's scientific research that suggests that there will be more turbulent flights in a climate-changing environment.

But the key point I often emphasize is there are things we can identify that are happening now, not 2080. These things are—and the question earlier from your colleague in southeast Texas—I think Adam handled it very well, the question about sort of this notion of uncertainties in reports. I don't think those reports are saying that things aren't happening. I think it's just semantics. We ran into that a little bit with the American Meteorological Society as well in terms of a report we issued because, as a scientist, we're trained not to say things in absolutes. And I think to our detriment to some degree, that's created some of the cloudiness, pun intended, in the discussion.

Mrs. FLETCHER. Yes, I'm a lawyer, and we also don't tend to speak in absolutes, but I do think it's an important point. And going back to that, another point you made—and I'd love to get anyone else's thoughts on this as well, so it's a sort of open to the panel—but the conversation about sort of climate change and whether humans have an impact seems to be a false choice. We know it's happening and it has been happening, and we also know that we can have an impact.

And so if anyone wants to weigh in on how we kind of navigate that messaging to say—especially for those of us—I think we have a lot of constituents who believe that it's happening, who see the effects, who want to understand that you don't have to choose between—there's not one scientific theory or another, that they work together. So Dr. Sobel, I know this is an area you researched. Maybe you can share your thoughts on that.

Dr. SOBEL. An analogy that I find useful in this is we know that humans die of natural causes every day, many, many people. And this has occurred since there have been human beings. But that would be—to say that humans die of natural causes would be a pretty weak defense in a murder trial. If I were accused of killing somebody, that wouldn't get me off. I would have to show why I didn't do it.

And similarly, I think the climate has always changed for many reasons, but we are a big reason now and we know that from many, many lines of evidence. So there's no inconsistency between those two things besides that analogy. There's many, many others one can come up with where there's multiple causes for things, but we know, you know, what the cause is now. So I don't know how to make the messaging better except to try to get that simple concept across.

Dr. SHEPHERD. And a point I often make is that trees fall naturally in the forest all of the time, but that doesn't mean chainsaws are hoaxes.

Mrs. FLETCHER. That is a great analogy. I love analogies. I guess the last sort of question I would put out there is just what do you all think—and there's very little time—but what do you all think is the most important thing that our constituents need to know about how to prepare for a future with these more intense hurricanes?

Dr. DONE. Yes, perhaps rather ominously is that the biggest signal is that the most intense events are changing the fastest. So the category 4, 5 hurricanes, evidence is building that we'll see more of them in the future. The heaviest rainfall events over Manhattan, for example, are going to be even worse in the future. So with that understanding I think it's time to act.

Mrs. FLETCHER. Thank you very much. I see I've gone over my time, and I yield back. Thank you all.

Chairwoman JOHNSON. Thank you very much. Mr. Waltz. Mr. WALTZ. Thank you, Madam Chairwoman.

Dr. Moore, as I'm sure we've discussed already, Hurricane Dorian, it was, as we all know, predicted to hit the east coast of Florida. I represent northeast Florida from Cape Canaveral up to Jacksonville. Three of the counties in my district, Volusia, Flagler, and St. John's, went under a mandatory evacuation based on the predictions. Obviously, Dorian never made landfall in Florida fortunately.

This was a huge dilemma and hugely frustrating of when our county officials make those calls to evacuate. You don't want to overact, but obviously—and to the tune of significant expenditures, but on the other hand you don't want to underact. And obviously this is a call that we all have to collectively make every time one of these storms threaten the United States.

In this case bottom line is we dodged—I don't even want to say we dodged a bullet; we dodged a missile. The category 5 storm could have been absolutely catastrophic. And at one point I think we were looking at Hurricane Andrew, Hurricane Michael, and Hurricane Katrina maybe combined in terms of the original track. I know I personally was hitting the refresh button on the predictions probably seemed like about every 5 minutes. I think NOAA and the National Hurricane Center did as great a job and as good a job as they could informing the public every 3 hours.

Can you just take a moment and discuss the primary differences between the models that you use, the European model that we hear so much about and NOAA's? Are there any data points you would recommend NOAA include or weigh more heavily to improve the accuracy of these models? It seems like in this case the European model was pretty accurate. And if you could just talk to that for a moment because there are literally millions of lives and treasure that are dependent—it's a great thing that we have it, but what can we do to help improve it?

Dr. MOORE. Let me make three points. First of all, to the emergency managers at NOAA that had to make that call, if anyone doubts why we made the call or that NOAA made the call, look at what happened to the island offshore where it did hit, and just project that damage into Florida. And if you had people in harm's way and not being evacuated, it would have been a missile strike. It would have been a missile strike.

Mr. WALTZ. Oh, absolutely. And just in the interest of time, I'm not questioning at all the calls that were made. How do we, as we look forward, improve the accuracy of our models?

Dr. MOORE. I think what we have to also do is look at how the Europeans conduct their research. They marshal the best talent across Europe, and they focus that talent really on one model. And they say this is where we're making our bet. We have tended to focus our talent across many models and, at the same time, we have only used a portion of our scientific talent. In particular the university community has not been in a position to be as engaged in the development of the next generation of numerical weather prediction.

And I hope with the EPIC initiative—and we thank Congress for this—and with the *Innovation Act*, that there is a recognition that the total United States scientific community from the social scientists to the physical sciences have got to be engaged in this grand enterprise.

Mr. WALTZ. Thank you very much. And I appreciate that answer. Again, in the few minutes I have remaining, and this is for anyone on the panel, Florida and Florida's Governor is, I think, leading the way in many ways in naming a Chief Science Officer and naming a Chief Resiliency Officer and trying to look at how we mitigate but then also how do we respond once we sustain that type of damage, and do we rebuild in smarter ways and more resilient ways than we have in the past? Would anyone care either for the time I have remaining or I'd appreciate it for the record of how we can be more resilient. I think Florida is leading the way post-Hurricane Andrew in wind resiliency, and we need to look at that in terms of flood resiliency with rising seas, with insurance markets, with property valuations. This is critical to the future of our State and the country. In the few seconds I have remaining, any responses, please?

Dr. SHEPHERD. I'll jump on it, but I think James has some thoughts as well. At the University of Georgia we've stood up an Institute for

Resilient Infrastructure Systems where we're thinking exactly about those because there is research by Phil Klotzbach and I believe—and Roger Pielke, Jr. and others that have shown the infrastructure along coastal regions has increased at the same time we're seeing these extreme events, so there has to be a sustained thought process for thinking about resilience and risk for our infrastructure in these regions.

Dr. DONE. Yes, just to build on that, just briefly, some work I did with my colleagues showed that the Florida building code is extremely effective against hurricane wind damage, and that's due to the strength of the code and how well it's enforced. And so building that nationwide would presumably bring similar benefits.

Mr. WALTZ. Thank you. And I welcome any other responses for the record. I think this is an incredibly useful conversation.

Dr. SHEPHERD. Water resiliency, too, and not just wind.

Mr. WALTZ. Absolutely.

Dr. SHEPHERD. Water gets underplayed, not just the surge——

Mr. WALTZ. That's right.

Dr. SHEPHERD [continuing]. The rainfall.

Mr. WALTZ. Thank you, sir. Thank you, Madam Chairwoman.

Chairwoman JOHNSON. Thank you very much. Mr. Lamb.

Mr. LAMB. And thank you to all the witnesses for coming, and I'm sort of picking up where you left off, Dr. Shepherd, because intense rainfall in short bursts has really increased in western Pennsylvania where I represent in the last couple of years. July 2018 was the wettest month on record for us. And many of the people that I represent live at the bottom of the big hillsides that line our rivers where we are. And many of these are old mill or manufacturing towns. That's why they were located there in the first place. And so these are middle-class communities.

I walked through a woman's home this summer who was born in that house, and she bought it from her parents and is in her 60s now and has just watched it flood every single summer for the last few years because, as Dr. Done helpfully pointed out in his written testimony, Pittsburgh has not really changed its stormwater infrastructure in the last century. We have failed to keep up. I place a lot of the blame for that at the national level and our inability to work together on an infrastructure package that actually meets the challenges of the modern day. So that's something we're not going to give up on and we're going to keep working on.

But I would like maybe if a couple of you could delve into the details of the scientific consensus that these intense rainstorms are caused by climate change and by global warming. You know, the average person on the street in Pittsburgh can tell that they're happening more frequently than they ever happened before. I have an Aunt Patsy who's 99 years old and lived at the bottom of one of those hills for a long time, and she's remarked at how different it is now.

But, Dr. Done I think it was referred in your testimony to a fundamental scientific principle that moisture in the atmosphere is increasing for every degree of warming.

But I think what would help me would be if you could describe how sure are we? You know, I've heard it said, for example, that the scientific consensus on anthropogenic climate change is now roughly equivalent to the scientific consensus on gravity. So is there another analogy like that that can help me communicate to my constituents how sure we are the cause of these rainstorms and the flooding that comes with them?

Dr. DONE. Yes, as a scientist, we look at two main drivers of extreme events. We look at so-called thermodynamic drivers, which is changes in heat and moisture, and then we look at changes in the circulation, so the changes in the weather patterns. We like to separate those two, and we're very confident in changes to the heat and the moisture—or changes due to those increases. We're less sure on the changes in extreme events due to circulation.

So when I'm talking about changes in extreme precipitation in the northeast, for example, I tend to talk about changes in frequency of weather patterns, you know, changes in high-pressure weather patterns or troughs of the eastern U.S. and talk about changes in the frequency of those different distributions. It seems to be an effective way of communicating extremes.

Dr. SHEPHERD. And I would say that the report that Adam and I both worked on for the National Academies, we ranked our confidence in changes in extreme events, and top 1 or 2 percent rain events we were very confident in the notion that they're coupled to climate change. Adam?

Dr. SOBEL. Yes, the influence of warming on extreme rain events is not quite at the level of gravity, but it's getting closer.

Mr. LAMB. Getting——

Dr. SOBEL. I mean, the rule that we used in the report that Dr. Shepherd just mentioned is that the more closely any given type of event is related to temperature, the better we understand it. And extreme rain events—so heat—that's why heat is the simplest, but extreme rain events are the next simplest because the water vapor in the air is so closely related to temperature.

The other thing I want to say here that's important to understand is that we have several ways of knowing what's happening in the atmosphere. One is the observations, the other is understanding the physics, and the third is the models that we use to predict climate and weather. In the case of extreme rain events, they're all giving the same answer at least in direction, so we understand there's more water vapor in the air, and the extreme rain events are the ones that are really good at squeezing that out. And so the way they change is closely coupled to the amount of water vapor in the air and thus to the warming. The models also predict they should be getting heavier uniformly. And we can see it in the observations. If we look at all extreme rain events——

Mr. LAMB. I'm going to cut you off there just because there was one last thing I wanted to get in before my——

Dr. SOBEL. Sure.

Mr. LAMB [continuing]. Time expires, but thank you.

Dr. Done, I think your work on the building codes in Florida is great, and I'd like to see more of it for different uses in parts of the country. It comes to mind for me that those two pictures you put in your testimony of Pittsburgh, the implication is the infrastructure hasn't changed.

There's one way in which the infrastructure did change in that time period you're talking about, which is that in the 1930s and 1940s America built a series of dams, locks and dams. But one of the dams in particular was a direct reaction to the flood of downtown Pittsburgh, the same place you have a picture of in 1936. And that dam cost roughly in today's dollars around $5 million to build. And there have been estimates that it's probably prevented over $500 million of flood-related damages that come from when you flood a downtown of a major city.

And so is there more work like what you've been doing that we could do from the government or from the National Science Foundation or otherwise? Can you just give me a little bit of insight into that?

Dr. DONE. Yes, that's a great point. Building dams is building resistance to extreme precipitation events. I think there's a lot to pursue in terms of building safe-to-fail infrastructure, so this would allow communities to absorb some of the event so rather than building a wall or a dam higher and higher and higher because there's always going to be a worse extreme, I think it's vital to— now we're in a new era of changing extremes to absorb this notion of what we call graceful failure, so where we absorb some of the shock but we can get back on our feet very quickly.

Mr. LAMB. Thank you. Madam Chairwoman, I yield back.

Chairwoman JOHNSON. Thank you very much. Mr. Posey.

Mr. POSEY. Thank you, Madam Chair, for holding this hearing today on forecasting and communicating extreme weather. I assure you it's very important to the citizens of Florida because it saves lives.

In 2004 and 2005 Florida was hit by a series of severe hurricanes, about five of them I believe, which caused damage in every single county. All 68 counties were damaged, and most of them pretty significantly. The insurance companies obviously went ballistic almost immediately and doubled or tripled everybody's rates because the reinsurers did the same thing. And we started getting copies of their gloom-and-doom forecasts for the future, that it would only get worse the next year and progressively and progressively and progressively. And surely that in just a few years there would be no insurance left for anybody or there would be no one who could afford the insurance.

And the State lawmakers were concerned by that, so they hired their own actuaries, their own data study people and the result was even worse than the insurance companies had told us, so it was pretty bad.

We received one unsolicited suggestion and analysis from a fellow named Dilley, a retired forecaster with the National Weather Service. Any of you ever heard of him? Just shake your head if you have heard of him before. And after he retired, he started tracking severe weather as a hobby. And so rather than go review the statistical maps and timelines and take a statistical approach to the whole thing, he focused on way upper weather patterns. And Mr. Dilley told us, he said I only go 8 years out with my forecasts, but he said I think you guys are good to go for about 8 years before you're going to have another severe hurricane hit Florida. And you may have a little action in the southwest part of the State, but it won't be severe.

Most people laughed at this guy, but history proved that he was right on the money. And so I'd like to know if you're familiar with his type of analyses, who does it, or other thoughts that you might have on it, and we might start with Mrs. Bostrom.

Dr. BOSTROM. Well, that is way out of my bailiwick, but I'd like to—something came to mind why you were talking, and that was the predictions of earthquakes. But you can sometimes be right about something even if the method that you're using isn't terribly good. So I would really need to know a lot more and otherwise conclude that this was a coincidence that he was correct.

Dr. SHEPHERD. Yes, I was going to say the same thing. You know, a broken clock is right twice I guess in a day and so——

Mr. POSEY. Well, he wasn't right once. He was right eight times, and everybody else was wrong eight times.

Dr. SHEPHERD. Sure. I mean, and that's still a small in the eight. So what I tend to sort of look at when I look at these sort of—because as meteorologists we hear all kinds of things about rodents that predict the weather and almanacs, but I tend to go with the sort of peer-reviewed literature. Now, I would certainly love to see his methodologies published, and I think that's what the peer-reviewed literature is for. So if Mr. Dilley is listening out there, I would invite him to sort of publish that methodology so it can be evaluated because he may be onto something. I would just suggest that I wouldn't want the National Weather Service and colleagues to sort of change their overarching principles and methods based on a sample of eight.

Mr. POSEY. I'm not suggesting that, just curious if anybody else had heard of that, you know, and what your thoughts about it are.

Dr. SHEPHERD. No——

Mr. POSEY. Dr. Moore, anybody else have any thoughts about it?

Nobody else thinks it's worth investigating?

Dr. SOBEL. I don't know this fellow's method, but we could talk about the reinsurance rates in Florida after 2005 if that's of interest. The reinsurance rates are influenced by a lot of things, and it is true that there's market forces—so there's an emotional response either to an event that just happened or a sequence of no events, so the rates go up a lot of there's a bunch of bad events, and they go down if there haven't been any for a while in a way that probably fluctuates more than it should.

The risk is sort of a slowly changing thing, but how often the risk is realized is a different thing. So you have car insurance, but you're not going to crash your car every day. So——

Mr. POSEY. Reinsurance——

Dr. SOBEL. Yes.

Mr. POSEY [continuing]. Is different in a lot of different ways.

Dr. SOBEL. Yes, but——

Mr. POSEY. Number one: They're all housed in the same offshore island, and they all charge the same rate and change their rates at the same time. We'd call that a monopoly if they were on the mainland.

I thank you, Madam Chair. I see my time is expired.

Chairwoman JOHNSON. Thank you very much. Mr. Foster.

Mr. FOSTER. Thank you, Madam Chairwoman. And thanks to our witnesses. And my apologies for having been in and out of this hearing.

One of the things I'm hearing a lot of excitement about from the computational physicists at Argonne Lab that I represent is the impact of artificial intelligence on things like dynamic mesh reconfiguration and getting effective models at small scales that are applicable to large-scale modeling. And they believe that in some applications that may buy more than a factor of 10 a maybe 100 in the amount of bang for your buck for a given level of computational power. And is that also happening in the weather prediction regime, and is there any way to quantify—is that going to buy us 1 or 2 days of additional forecast accuracy or something along those lines?

Dr. DONE. Yes, there's a great methodology. And, you know, at the National Center for Atmospheric Research, we've engaged with this so-called mesh refinement technique with the weather research and forecasting model, so we have about a decade of experience in real-time forecasts and hurricanes. Now, the key point is that they're computationally cheap, and so we can run it more than once. We can run, say, an ensemble of 50, so this allows us to really describe and draw out the future possible scenarios to enable more robust responses.

Mr. FOSTER. Yes. But do you have any estimate at this point of what that's going to buy in terms of accuracy? You know, because you sort of hit this chaotic wall at a week or 2, and I was wondering does this allow you to push closer to the chaotic wall or is the chaotic wall so sharp that even these techniques aren't going to make things better? Yes, Dr. Moore?

Dr. MOORE. The chaotic wall is there irregardless, but what this allows you to get at is the so-called convective resolving models so that the models actually can begin to resolve clouds and convective—particularly the third dimensional motion of the atmosphere. And that is extremely important in any type of numerical weather prediction. The work that is being done at Argonne really is first rate in getting us toward the capability of doing convective resolving. And then you can begin to assimilate satellite data at these higher spatial temporal resolutions to further tamp down the chaotic aspect. Satellite observations and sunsets always draw us back to truth.

Mr. FOSTER. And then at the end of the rainbow is, you know, the concept of being able to actually steer hurricanes. And, you know, about a decade ago Bill Gates had this patent that got a certain amount of press that, you know, with a relatively small change to the surface temperature of water the thought was you could either suppress or steer hurricanes. And has that been looked at and modeled by anyone? Is it completely hopeless or if you could do something relatively minor, can you actually steer hurricanes?

Dr. SOBEL. I don't think you can do anything with something minor. Hurricanes are very enormous and have a huge amount of energy. So steering I don't think is practical. What could be done——

Mr. FOSTER. Well, what I was thinking of is if you're near a——

Dr. SOBEL. Yes.

Mr. FOSTER [continuing]. Disturbance in the air, the chaotic wall—

Dr. SOBEL. Yes.

Mr. FOSTER [continuing]. Would have a very high impact if you could calculate what that impact would be, you know, the butterfly——

Dr. SOBEL. Oh, so you're not talking about actually changing the path of the storm but just predicting it?

Mr. FOSTER. Well, if you could predict it accurately and predict the effect of a small perturbation early in the development of the storm——

Dr. SOBEL. Right.

Mr. FOSTER [continuing]. Conceivably, that would have a large impact on the trajectory.

Dr. SOBEL. Yes, I mean, it may be physically possible. My guess would be that you still take a huge amount of energy to do it. What I think is better understood and what I think the Gates project was getting at was changing the intensity, which if you can cool the ocean's——

Mr. FOSTER. The surface temperature, right.

Dr. SOBEL [continuing]. Surface enough you could do that, but it still would take—you'd have to do that over a very, very large scale and have to operationalize it very quickly as the hurricane developed. So my sense is that the cost it would take to do that would be better spent on measures to protect life and property and get people out of harm's way. But that's——

Mr. FOSTER. Well, I was wondering if something that can be done very early in the development may have a very large lever arm to affect the course and, you know, things like, I don't know, seeding clouds and stuff that I guess it's sort of well-understood and pretty minor might be able to, if it's done 2 weeks in advance, actually have a noticeable and useful effect.

Anyway, if there's any work that's been done on that, that'd be fascinating to read myself to sleep with it when it's possible because it's not obviously from first principles impossible.

Let's see—and I have now 15 seconds, yes, but—so I have no 15-second questions left, so I'll yield back.

Mr. BEYER [presiding]. Great. The Chair recognizes the gentleman from Illinois, Mr. Lipinski.

Mr. LIPINSKI. Thank you, Mr. Chairman. I thank all of our witnesses for testifying today.

I have long been on this Committee, have been concerned about staffing at the National Weather Service, and the Administration proposed eliminating 20 percent of the forecasters, closed some of the offices on nights and weekends has been very concerning to me.

I want to ask a question—I know no one is here from the National Weather Service, but I just wanted to ask this question and see if anyone had experience or comments on this. I know after major storm events that lead to major economic impact and fatalities, the National Weather Service employees are called in to evaluate response performance. So I'm curious if any of the witnesses on our panel have reviewed these service assessments and used them in your research efforts, particularly as you evaluate impacts and responses to extreme weather events. And if so, have you noted any impact from the understaffing that impedes National Weather Service's ability to respond during extreme weather events? Dr. Shepherd?

Dr. SHEPHERD. I haven't done research in that, but I will say that it's important that our National Weather Service offices are fully staffed, particularly in these extreme events or these sort of long-term sustained events. I've heard stories of National Weather Service employees during shutdowns or sequestrations and things like that slogging to work in snow and those types of things, and those aren't hyperbole. Those things happen.

I wrote an article once about the sort of psychological fatigue and the sort of mental aspects that meteorologists deal with in these offices. You know, the first responders, kudos to them and thank them for what they do in the emergency response, but oftentimes meteorologists carry a burden because they are forecasting and predicting events that are going to change people's lives. And that's a tough sort of psychological burden.

So I say all of that to say that again, I have not done research— I don't know if anyone has—but I would always advocate that our National Weather Service offices are never cut in terms of staffing. In fact, if anything, they probably need to be upgraded because they've been short-staffed in some regards. I know my friend Louis Uccellini at the National Weather Service has tried to be responsive to this, and there are pressures and forces beyond his control, but those offices need to be fully staffed.

Dr. MOORE. Let me just mention, we've been critical of numerical weather prediction capability of our models and so forth, but that should not extend into being critical of the National Weather Service per se. The work that they do—and in particular I remember the Storm Prediction Center at the University of Oklahoma National Weather Center talking about a tornadic outbreak in the Illinois-Indiana area in November. And they were speaking about it 10 days out. And eventually we got into the endgame and there was a tornadic event, but by that time FEMA (Federal Emergency Management Agency) had facilities on the ground, the body politic was prepared. And really the fact of what we're able to do in terms of weather prediction and protection for our society is really extraordinary. It really is one of the grand accomplishments of science.

Mr. LIPINSKI. And Dr. Bostrom?

Dr. BOSTROM. Yes, I'd like to echo what Dr. Moore said and also what Dr. Shepherd said that the National Weather Service does an enormous service to the country in protecting people, and there have been in the testimonies earlier this year from Dr. Uccellini and others examples of how the Weather Service has been able to provide information that has equipped emergency managers to help their communities.

Phaedra Daipha has written some research that is ethnographic that look at what goes on in Weather Service offices, and you can see from that that they're overtaxed, that they're underequipped and understaffed. And they work really hard to cover their jobs and are doing the best they can. And her work is a call for better staffing at the National Weather Service. And Dr. Uccellini and others have tried very hard to make sure that the staffing has improved. They've made some progress, but, as you can tell, it's an uphill slog.

I have not done research on the service assessments, but they're a really important contribution in terms of providing feedback on how this all works. And it's been clear from the service assessments that have been done that there are a lot of things that need to be improved about how the whole system works. That does not mean that people are not doing their jobs. They're doing an excellent job.

Mr. LIPINSKI. Thank you. And I don't really have time for a question, but I just wanted to point out the recent assessment of impacts of climate change on the Great Lakes put out by the Environmental Policy and Law Center. And this is something that I think really brought a lot of attention to the impacts in the midwest— I don't have time right now to ask about that, but maybe in followup I will ask you all about that. But thank you very much. I yield back.

Mr. BEYER. Thank you. And the Chair now recognizes the gentlelady from Virginia, Ms. Wexton.

Ms. WEXTON. Thank you, Mr. Chairman. And thank you to the panelists for coming and joining us today.

The frequency and severity of extreme weather is clearly something that's of interest to everybody on this Committee. I don't think there's a single one of us who our district has not been impacted in some way. And my home, as the Chairman stated, is in Virginia where we have seen increased inland and coastal flooding, more days with the heat index above 105 degrees, and lots of these extreme rainfall events that have caused a great deal of damage to roads and infrastructure. And, in fact, in the lead up to Hurricane Florence, almost a quarter million Virginians were ordered to be evacuated from low-lying flood-prone areas throughout the Commonwealth. So we rely as a State and our localities as well rely on NOAA and the National Weather Service to help them make informed decisions about when the public safety is at risk.

So I would ask of everybody on the panel, how important is it for the Weather Service to be federally funded, readily available to the public, and completely separate from private interests? Dr. Shepherd?

Dr. SHEPHERD. I think this is an area where we've come a long way in the weather enterprise. I'm speaking sort of on behalf of the weather enterprise. There was I would say a tension between the sort of public-private partnership. But I think we found our rhythm. I think there are very clear roles for both the private sector and the public sector in this partnership. I think the recent weather legislation has sort of provided other opportunities for the private sector to kind of get into this.

I still view many of the services provided by the National Weather Service as sort of what I consider like police and fire services, sort of federally sort of designated services that we need, irrespective of profit margins, irrespective of other things because they provide very critical information. But I think in terms of the value-added services, some of the nimbleness and some of the observations, I think there is a role for the private sector as well. But I think some core services certainly need to be maintained. I think EPIC is providing a nice model for us, though, of how a public-private partnership can work.

Dr. DONE. Yes, I'd like to speak to the value of sustained National Weather Service staffing. Through decades of observing weather, you can really understand and get some unique intelligence on how the weather works.

So I'll give a quick example that happened just 2 weeks ago over in northern California. There was an outbreak of so-called dry thunderstorms. So it was spoken a lot today about the advances we've seen in our simulation capacity, but these kind of events, they don't exist in our weather prediction models, so it's only through experience that the forecasters could see the weather pattern, they know that there's a risk of these so-called dry thunderstorms and the lightning that can trigger or raise fire risks. So this speaks to the importance of a sustained staffing of the Weather Service.

Ms. WEXTON. OK. Before we go down the rest of the panel, I just want to express to you a concern that I have because the gentleman who has been tapped to lead NOAA, Barry Myers, has previously advocated for NOAA to restrict the amount of weather information that is provided to the public. And he uncoincidentally at least was one of the majority shareholders in a private company AccuWeather. And even though they get all of their information from the National Weather Service and NOAA, they then want to sell it back to the American public at a profit. And I mean, there's inherent conflict of interest in there in my mind.

So I guess, you know, building on my earlier question, do you see potential conflicts when you have private industry who is restricting the information that's getting out to the public in weather? Dr. Sobel?

Dr. SOBEL. So I think the private sector has an important role to play, but as Dr. Shepherd said and as I think you're eluding to, the Federal Government does the core work to make the weather forecast possible, launches the balloons and runs the satellites, and runs the models. The private sector is doing value-added on top of that. So it is like police and fire. There should be a public weather forecast available to everyone done by the government employees, and the private sector has a tremendous role and is doing very well, adding value to that, and so I think the way it's been done historically has been working, has been evolving in a way, but we should keep the National Weather Service strong and making public forecasts as it has been.

Ms. WEXTON. And I have no qualms with the private sector adding value to that work which is provided by the National Weather Service and NOAA. What my issue is, restricting——

Dr. SOBEL. Yes.

Ms. WEXTON [continuing]. The information that is readily available to the public for the purposes of getting a profit so——

Dr. SOBEL. I completely share your concern there.

Dr. SHEPHERD. Yes, and the taxpayers pay for these services, so I will just say without commenting on any specific person or their thoughts or companies, but, you know, I often remind people that the satellites and the weather models were paid for with their taxes.

And by the way, the National Weather Service is funded on about the cost of a cup of coffee for every American citizen, so it's one of the biggest values in the Federal Government.

Ms. WEXTON. It certainly is. And on that note I don't think we can improve on that testimony. And I see my time is expired, so I'll yield back. Thanks.

Mr. BEYER. Thank you, Ms. Wexton. I now yield 5 minutes to myself for questions.

I'd like to just begin by thanking you very much for making an extraordinarily powerful case for the urgent need for all of us to address climate change and in any way that we can, which leads me inevitably to conversations about carbon pricing because I think most of us would agree, you know, every economist across the political spectrum from very conservative, very liberal thinks that we should price the things we don't want as high as we can and the things we do want as low as we can.

In fact, it's been very bipartisan. Francis Rooney, who asked some great questions earlier, is a lead sponsor of a carbon pricing bill. And Carlos Curbelo, recently of this body led that effort significantly. And we have all these corporations, ExxonMobil, BP, I think Chevron or Shell, you know, the world is coming together on it. But my sense is that this Congress won't take up a carbon pricing bill because we're not going to get anywhere in the Senate with it. And we still have some political—actually partisan resistance.

So I want to go back to the NOAA estimate from earlier this year that we had $91 billion in damages in 2018 from extreme weather events and $306 billion in the U.S. from extreme weather events. When you put all those together, doesn't it sound like we have a hell of a carbon fee anyway? Dr. Done?

Dr. DONE. Yes, absolutely. That's a very important point. We are already paying for climate change through our losses. And in fact I know—and through my conversations with the reinsurance industry that they can see the footprints of climate change in their loss data. So climate change is becoming central to everything—well, much of what the reinsurance industry does. They've realized they can no longer look backward if they want to assess today's risks; they must also look forward. And that demands scientific understanding of what's tomorrow, 10 years, 50 years from now.

Mr. BEYER. To that exact thing, Zurich Insurance came to see me and my wonderful staff recently to make exactly that case, that is a look at how much reserves they have to have down the road, please do something about climate and start with carbon pricing.

Dr. Shepherd, the compound extreme weather events, you've talked about that a little bit, the notion that, you know, the hurricanes, followed by the flood, the wildfire, the heat is followed by the wildfire. Can you talk about how difficult it is, what progress we can make in terms of predicting the compound weather events?

Dr. SHEPHERD. So this conversation we were having earlier— thank you for the question. I think that there's inherently nothing different about the model's ability to handle a singular event from a compounded event, but as we were talking earlier, those secondary events, the complexities of those, how they may be resolved—I think Professor Moore has been talking about these— there may be some secondary events that are not as well-resolved that are underplayed. We just saw that, for example, with Imelda in terms of that trough, that mid-latitude, non-tropical system sort of suffering some.

So I think if there is one area among many actually that I would recommend further study, research funding, et cetera, through the National Science Foundation, NOAA, NASA, or others is in understanding how our modeling capabilities are addressing these multiple compound events because I think this is an emerging area of study. I know Professor Sobel mentioned that there is some work in a workshop recently that has been held. But I think this is fertile ground for new research.

Mr. BEYER. Yes, thank you. And Dr. Bostrom, Dr. Done had talked about the difficulty of attribution, especially when you look at so much, you know, impervious surfaces everywhere, the urbanization of the world, a growing population, 7 billion. How do we tease out the climate change signal from the other factors that are affecting the extreme weather events and the impacts of the extreme weather events?

Dr. BOSTROM. As in any scientific problem, what we try to do is come up with counterfactuals and then figure out what the specific attribution can be for any specific factor that's driving a change like climate change. So as everybody's been talking about, we have models of what's driving specific damages, what's driving specific hazards. And in those models we're generally using simulation models to try to find out what we can attribute to a specific event.

That's hard to talk about, and so I think the easiest thing for people to understand is that—to tell them—to show them examples where they wouldn't have seen the same kind of damage or the same type of effect without the specific fingerprint or thumbprint of climate change.

Mr. BEYER. Great. Thank you all very much. We have been abandoned by the other Members of Congress, who are off doing the people's work, so I just want to thank you very much. And thank you for all that you do in your professional lives to keep us safe, to work on our economic and our personal security. We're very, very grateful for the work that you do.

And thanks for being here today. It's often an anxious event appearing before all the mean people on this dais, but you survived it very well.

The record will remain open for 2 weeks for additional statements from Members and for any additional questions the Committee Members may have for you.

So with that, the witnesses are excused. The hearing is now adjourned.

[Whereupon, at 12:10 p.m., the Committee was adjourned.]

APPENDIX I: ANSWERS TO POST-HEARING QUESTIONS

Responses by Dr. J. Marshall Shepherd

Responses to Questions from Chairwoman Johnson

R1: The most important aspect of EPIC in my mind is that it will allow researchers from multiple sectors to work on a common code/software for our large-scale weather model. In the "old way," new ideas struggled to get into the system. This should accelerate knowledge transfer from the private, federal, and academic sector and remove "silos." Increasingly, these sectors of the weather enterprise have been working together under the auspices of the American Meteorological Society (AMS) **Commission on the Weather, Water, and Climate Enterprise (CWWCE).** I am the former President of the AMS, and I think this professional society and this Commission could play a key role as an organizing principle for EPIC's governance structure. The AMS website explicitly says that CWWCE "*is charged with developing and implementing programs that address the needs and concerns of all sectors of the weather, water and climate enterprise; promote a sense of community between government entities, private sector organizations; and universities; foster synergistic linkages between and among sectors; entrain and educate user communities on the value of weather and climate information; educate policy makers on the value and operations of the enterprise; and provide appropriate venues and opportunities for communications that foster frank, open and balanced discussions on points of contention and concern. In other words, the "prime directive" of the CWWCE is to engage the government, academic, and private sectors on pressing and strategic issues on behalf of the Society. These interactions may extend to involve other AMS Commissions, the user community, and disciplines beyond that may request assistance with enterprise communication, growth and development matters.*" UCAR is another trusted integrator within our community that I trust to keep the players on the field and talking. Even though it has strongest ties to the academic sector, it is also a trusted source at the table. AMS operates across all sectors and is trusted by all constituents within our field. I do believe, however, that some core responsibilities must remain under federal auspices beyond the research and development functions. I still view ownership and delivery of weather model, forecast, and observation data/functionality as a core function for government. Additionally, there must be a significant level of security for an open-source approach like EPIC. Resources must be provided to allow contributors to be functional, yet the core software must be secure.

R2: Compound extreme weather events are emerging as a major concern. Research initiatives are needed within the portfolios of NASA, NOAA, and NSF to advance understanding, data collection, and prediction. For example, the remnants of Tropical Storm Imelda caused significant flood impacts to Texas in 2019. It was a tropical system compounded by a mid-latitude weather system approaching from the west and stalled motion. Current models may not always capture the coupled atmosphere-ocean-land processes associated with these systems due to deficiencies in computing resources, lack of observations at proper scales, and poor understanding of the mechanisms. The most important needs are: 1. Funding programs explicitly addressing compound events, 2. Improved regional and state mesoscale observations via meso-networks or MESONETS (with 4-dimensional capacity), 3. Continued support of advanced satellite-based observations and implementation of the NASA decadal survey, and 4. Adequate computing resources to accommodate more sophisticated data assimilation schemes and volume.

R3: One of the most latent aspects of climate change and extreme events is the disproportionate effects on marginalized or vulnerable populations. I have recently discussed what I call the "weather-climate gap" (Recent TedX Talk on this by me in Washington, DC: https://www.youtube.com/watch?v=IYEm3eyDHrk).

When a heatwave or hurricane affects a population (rural or urban), citizens at the lowest end of the income gap have the greatest sensitivity and least amount of resiliency. A similar narrative has long been advanced in the environmental justice field, but climate or weather justice studies are lacking. Cross-disciplinary and inter-disciplinary research are required to understand the drivers and consequences of the weather-climate gap. I also believe that economic and social policy (e.g., housing, health care, jobs) must increasingly account for the weather-climate gap. We can no longer view this as an isolated fringe or environmental issue. It is strongly tied to the very fabric of domestic policy. Events like Hurricane Harvey, Hurricane Maria, or heatwaves further amplify poverty, health issues, access to food, water availability and well-being. And ironically, the most vulnerable people within our population often carry the lowest carbon footprints. This mismatch also deserves thought at the policy level, particularly as carbon taxation and other pricing strategies emerge.

R4: Sub-seasonal to seasonal (S2S) prediction is important because it carries implications for understanding things like El Nino and La Nina. While there arc many other atmospheric-oceanic modes (NAO, AO, MJO, PDO, etc.), these two have entered the public sphere. All of them influence hurricane activity, drought, wildfires, and temperature extremes. Better S2S prediction, enables real possibilities in planning, prevention, and risk management on seasonal scales. Computing resources and proper allocation of access to such resources remain challenges as do fundamental research questions about the relationships to extreme weather events. A recent paper by Vitart and Robertson (2018 in Nature, https://www.nature.com/articles/s41612-018-0013-0) argues "S2S forecasts of extreme events could be integrated into a "ready-set-go" framework between seasonal and medium range forecasts, by providing an early warning of an extreme event a few weeks in advance. Finally, S2S forecasts can also be used to investigate the causality of some extreme events..." NOAA's S2S (https://cpo.noaa.gov/Meet-the-Divisions/Earth-System-Science-and-Modeling/MAPPIMAPP-Task-Forces/S2S-Prediction-Task-Force) task force wrapped up its work in 2019, but perhaps it needs to be extended.

Responses to Questions from Mr. Walz

R1: This question could be answered in a multitude of ways, but I would simply offer that we are thinking about these very solutions at the Institute for Resilient Infrastructure Systems (IRIS) at the University of Georgia. We recently joined the Natural Infrastructure Initiative (Nil), a team of private companies and organizations promoting adoption of natural infrastructure in the United States. Natural infrastructure approaches offer real potential going forward and should be on the radar of Congress. if not already.

I think we also need to stop thinking of "resiliency" as bouncing back to the way it was. The new realities of climate change and extreme weather will require "reimagining" infrastructure and policy. I am a part of IRIS and its director, Dr. Brian Bledsoe, has said, "I don't think of resilience in the traditional sense, in cutting how long it takes to tum the lights back on....resilience is seizing an opportunity to move into a state of greater adaptability and preparedness - not just going back to the status quo." Florida already has post-disaster development plans at the state and community level so this places your state in an advantageous position for rapid recovery from a hurricane or flood. Should all states require these plans? Perhaps. Additionally, even Florida's plans should be dynamic and allow for "new norm" thinking rather than let's just "recover and rebuild."

I actually believe that all communities should have also have climate resiliency plans. There are some models within coastal communities of Georgia, for example. These plans must include flood or inundation maps based on realistic climate projections as well as return-period updates for floods that represent the new climate realities.

The National Flood Insurance Program also needs a permanent fix not continued band-aids.

Responses to Questions from Ms. Horn

R1: I would be very supportive of a national assessment of weather-related risk. I know that NOAA's NCEI tracks costs associated with weather events as do many private risk or insurance organizations. There is also some mention of these issues in the National Climate Assessment report. It would be prudent to have as much information as possible from credible sources and on a routine basis (perhaps every 4 years).

R2: These plans must include flood or inundation maps based on realistic climate projections as well as return-period updates for floods that represent the new climate realities. We fundamentally have a flawed way of communicating flood risks. Almost everyone misinterprets what a "500-year flood means," especially as climate change throws frequency and intensity out of "whack." People fear tornadoes and hurricanes but take a more relaxed approach to rainfall and flooding. There is a perception that rainfall flooding is common. People also misjudge flood risks because of "optimism bias" and the notion that their immediate concern (getting to their child in daycare, for example) outweighs the perceived risk of driving through a flooded roadway. Cute slogans aren't enough. One of my former graduate students at the University of Georgia and colleagues have proposed a flood index that captures the attention of the public the way the EF or Saffir-Simpson scale does for tornadoes or hurricanes. It is a challenging research-to-operations problem (R20), but I a problem that can be solved. We must also continue to support social science research at the intersection of weather and climate. Such studies help us to understand how and why people respond to flood warnings differently. From that type of research, more targeted messaging or policy can evolve.

Generally, I believe the National Flood Insurance Program also needs a permanent fix not continued band-aids.

R3: See previous responses to Chairwoman Johnson

Responses to Questions from Mr. Lipinski

R1: The Midwest/Great Lake region will be vulnerable on many fronts.

- Exposure to increased summer heat and heatwaves reminiscent of the deadly Chicago heatwave a couple of decades ago. I particularly worry about vulnerable populations (minorities, elderly, young children) with inadequate means to adapt to the new realities of climate change (see also my response to Chairwoman Johnson).

- Increased variability of rainfall and drought critical to Midwestern agricultural productivity. There is evidence that the trends in the heaviest rain events (top 1-2%) are increasing in the Midwest. This means that given current infrastructure assumptions, flood risks will increase, particularly in urban areas. More rainfall also means more sewage and fertilizer discharge into the Great Lakes.
- Possible increased extreme heat AND cold events as the jet stream becomes more amplified because of the "Arctic amplification" effect. This would leaded to the aforementioned extreme heat events but also more extreme cold/snowstorm/blizzard events. The work of Dr. Jennifer Francis at Rutgers University should be consulted for more about Arctic amplification and connections to the jet stream.
- Great Lakes temperatures will continue to increase, and this creates challenges for ecosystems. For example, {would expect an uptick in harmful algal blooms and fish kills due to hypoxia.
- Increased evaporation rates in a warming climate could cause falling water levels. Sea ice cover may also decrease. These could have mixed effects on commerce, transportation, and tourism. However, studies that I have reviewed suggest that an overall negative impact for the region.

Responses by Dr. James Done

Answers to Questions for the Record Submitted by Members of the House Committee on Science, Space and Technology

Answers to Questions Submitted by Chairwoman Eddie Bernice Johnson

1. How can the EPIC program foster collaboration between the public, academic, and private sectors of the weather enterprise? What role can and should the academic and research community

play in the development of the governance, goals, and implementation of the EPIC program?

Through a community model activity, the government can bring to bear private and academic sector resources and talent by providing a platform where all can mutually contribute to model improvements and share in their rewards. When governed correctly, a government community model behooves the private sector to utilize the superior numerical weather prediction gained through academic and government research and data assimilation achievements to improve their tailored products for their customers; it entices the academic sector to utilize the platform to conduct peer-reviewed science to push the boundaries of our understanding of the Earth System. *A* well governed community modeling system also naturally pairs well with a well governed research to operations system. Sustained two-way interaction across research and operations is the key to accelerating advances in understanding and forecasting that are aligned with community needs.

It is critical that the government create an academic-styled community modeling system because the academic sector is the only one that has achieved a well-used, well-governed, open, and effective modeling paradigm. The government, because of the necessarily closed nature of operational model management, lacks the culture, tools, and talent to perform this well independently. The academic community should have an outsized role in the development of the governance, goals, and implementation of the EPIC program, leveraging best practices and lessons learned. The U.S. is uniquely positioned for success owing to the breadth and depth of academic expertise unmatched by any other country or union of countries.

Additionally, the nation needs to accelerate the transfer of breakthroughs in research into operations in a more efficient and effective way. By tapping the breadth and depth of the US academic enterprise with its expertise that no other country or union of countries can match we will be much better positioned to drive the latest research advances into operations in a systemic way. This is a key ingredient that is missing and will enable advances to be implemented in a timely and scientifically tested manner.

2. To what extent can current models predict and forecast correlated or compound extreme weather events? What are the gaps in the research to understand the underlying physical processes, and how could federal funding help to address those gaps?

Extreme weather events such as flooding rains and damaging winds are of paramount societal significance. Yet a sequence or co-occurrence of events can cause societal impacts far higher than from individual events alone. Especially severe impacts can arise from connected events straining risk management practices that are often predicated upon the independence of extremes. The full range of physical mechanisms that connect extreme events has only just begun to be explored. Linkages may arise through local causal chains or via large-scale interactions across earth system components such as the oceans, land surface, and cryosphere. Moreover, improved understanding of the processes that connect extreme events has the potential to bring near-term advances in sub-seasonal to seasonal (2-weeks to 2 months) forecast capacity for extreme events.

The topic of correlated or compound extremes was identified as a key research need in the Climate Science Special Report of the Fourth National Climate Assessment. While there has been a surge in research focused on connected extremes over the past decade, much of it has focused on statistical characterizations of concurrent co-located extremes such as extreme wind and rain, or extreme heat and drought, rather than on the physical processes connecting events and their predictability.

Federal funding can help address these research gaps by supporting fundamental science advances through research grants, maintenance and expansion of extreme weather datasets, and continued support for the computational infrastructure necessary to interrogate big data. Federal funding is also needed to facilitate interactions between researchers and our communities to understand how communities are impacted by correlated or compound extremes and ensure the science is aligned with societal need.

3. What steps has the insurance industry taken to incorporate climate change information into their risk calculations?
   a) How can we ensure that premiums remain affordable? Is it possible that eventually some areas in the U.S. will become uninsurable?

Climate change has become a key concern for the reinsurance industry. Their world was rocked by Hurricane Andrew that hit Miami, FL in 1992. This event drove losses twice as high as anything that came before. They realized they can no longer look backwards to understand today's risk. They must also look forwards. Since then the industry has actively pursued solutions to our changing extreme weather that reduces impacts and speeds recovery.

They are now developing world-class risk models that incorporate effects of climate change. My experience as a Willis Research Fellow, where I collaborate with the reinsurance industry, taught me that there are huge potential gains in our ability to protect life and property through deeper integration of climate change science and risk management.

Insurance is a key tool to spread risk and to get help to where it is needed fast. A key concern is in keeping insurance affordable in a changing climate. One approach is to combine new insurance structures with climate change adaptations. Building codes, for example, have been shown to reduce hurricane wind losses to homes in Florida by up to 70% compared to homes not built to code. This adaptation reduces the absolute cost of the risk, and in turn keeps insurance premiums affordable.

The potential for some regions of the U.S. becoming uninsurable would have significant implications for the vulnerability and viability of local populations. A key requirement for a region to be insurable is a robust understanding of the hazard risk. This understanding is typically based on past events. But where the historical record is short, or the events particularly rare, such as hurricanes in the Northeast U.S., state-of-the-science computer model simulations are needed to increase our understanding.

4. Why do we need a decadal study on the terrestrial U.S. weather enterprise?

The National Academies of Sciences, Engineering, and Medicine perform decadal surveys that are vital for scientific progress in a number of disciplines. The weather community itself has benefited substantially from the Earth Observations decadal (which includes NOAA, NASA and USGS), but these benefits have focused exclusively on the capacity of data remotely sensed from space. To ensure that the Earth System Science community can appreciate the full potential of improving our understanding of the whole earth system (atmosphere, ocean, land, and cryosphere), we need a decadal survey that evaluates the future potential of the relevant science being conducted as well as the earth system data we currently produce. The survey is needed to evaluate the future potential of forecasting from days out to seasons, where are the computational technology gaps, and how extreme weather risk is communicated. This study provides the foundational understanding to understand how the weather enterprise (academia, public and private) can best adapt and support meaningful advances in our ability to enhance the protection of life and property. The urgency of this idea is particularly pressing in the new era of the internet of things era, where every platform has the potential to be a sensor or carry one that adds value to the forecast.

A decadal study is a study of national capabilities and will require coordination across federal agencies and engagement in a sustained process that transcends any one administration. It is not a restructuring or a "once in a while" national academy study but a process that enhances the coordination of community academic, private sector and agency officials to identity priorities. It is a continuous process that seeks continuous improvement and engages all players in the weather enterprise with a realization that they each bring different capabilities. It also recommends those programs that are no longer needed and can serve as a valued 3rd party set of eyes on the enterprise. This long-term study is not merely a collection of one-off studies, but directs the weather enterprise towards success.

As examples, a decadal would assess if EPIC is indeed on the right track and offer suggestions for improvements, would examine our nation's approach to seasonal to sub seasonal forecasting and would work to develop what the observational requirements would look like at a national level to ensure continued progress in improving forecast skill with a seamless approach from observations, to data assimilation and end result modeling that continually improves with metrics for improvement that policy makers understand.

5. How can we improve the research to operations [R20} and operations to research (02R} process when it comes to extreme weather? How has UCAR/NCAR successfully navigated this R20 and 02R process with NOAA previously?

Fostering and enabling many R202R pathways across the entire weather enterprise will yield more improvements and uncover efficient business models for further success. While there are great opportunities to leverage the research performed by OAR to improve NOAA operations, the government should not limit itself to this solo pathway. The academic community and the private sector are willing and excited to contribute to R202R endeavors at NOAA. Similarly, the private sector and academic community already interact considerably to improve private sector products and services and provide hands-on research opportunities for students.

An Example of UCAR/NCAR successful navigation of the R202R process at NOAA:

NSF invested in the creation and operation of the Weather Research Forecast model, a limited-area model utilized by the weather research community for decades through community modeling governance at NCAR. NCAR scientists configured WRF in a manner where specific Earth System conditions could be forecasted through supplementary code packages. Other NCAR scientists created a hydrology code package that forecasted river and stream levels across parameterized areas. NCAR worked with the private sector- Barron Meteorological Services- to operationalize this WRF-Hydro model in Romania and Israel with significant success in forecast hydrology. When NOAA recognized the capacity of WRF-Hydro, it operationalized it nation-wide as the National Water Model. By leveraging the successful research and operationalization of new technologies by the academic and private sectors, NOAA is now able to provide forecast hydrology products and services to the nation.

More recently, NOAA's National Water Center in Tuscaloosa, Alabama asked NCAR to use the WRF Hydro model to create the new National Water Model. This effort came in under budget and ahead of schedule. The RZO success was predicated on the clear need to do better in the area of water forecasting. Part of the reason this RZO worked well is NOAA had no internal capability to develop hydrologic models (hence no internal research pushback on extramural approach) and had to go to the research ecosystem and find the best approaches. The lack of "antibodies" within NOAA in regards to hydrologic modeling is the very antithesis of weather modeling where there are many competing schools with different approaches. The point here is critical; let everyone compete and let the best research win. Internal NOAA pressure to show the value of its research can lead to poor choices for the nation.

6. What are the societal benefits of a national dataset of extreme weather impacts?

Our understanding of how society is impacted by extreme weather events is grounded by what has happened in the past. Scientists and risk managers analyze these historical events to understand how we have been impacted in the past by, for example, hurricanes, severe thunderstorms and winter storms. Scientists also use these records to develop impact models. These models may then be used to simulate historical events and also generate 'what if scenarios. What if a category 5 hurricane hits Tampa, FL? What if the Florida building code was rolled out across the nation?

Critically, these models are only as good as the underlying data. Existing available datasets of historical impacts either have long records or are at a high level of detail, but not both. There is an urgent need for cooperation between the insurance industry, emergency managers and the federal government to understand how to pool extreme event data and curate detailed impact datasets of historical events, and for events going forwards. These foundational datasets will become a national asset. They will accelerate our understanding of extreme weather impacts, and ultimately enhance protection against extreme events.

Answer to the Question Submitted by Mr. Michael Waltz

1. Florida is leading the country in resiliency and mitigation in many ways, including the wind standard in the state building code. How can Florida and the United States be mare resilient to flooding and haw should we rebuild infrastructure more resiliently?

The Florida Building Code reduces hurricane wind losses to homes by up to 72% compared to homes not built to code. The code is also cost effective, giving 2 to 8 dollars back in mitigated losses for every dollar spent building to code. The Florida Building Code serves as a leading example of effective public policy for other states to consider across a range of extreme weather phenomena.

We live in an era of rapidly changing flood risk. Rising populations are combining with rising rain rates to elevate flood potential above what we've seen in recorded history. Take the case of Houston. All the ingredients for high flood risk converge on Houston. Frenzied development is occurring in a region that is already prone to floods. Now climate change is intensifying rain rates. Indeed, NOAA recently quadrupled the likelihood of extreme rainfall for the Houston area.

It's essential for the science to continue to empower communities. Recent advances in flood plain mapping are providing more comprehensive views of our flood risk. These new views find $5.5 trillion of assets lie within our 100-year flood plains. And we know that the 100-year flood has about a 25% likelihood of occurring within the typical 30-year mortgage.

And rain rates are rising. There is mounting evidence that the heaviest rain rates will rise by about 4% for every degree Fahrenheit warming of our atmosphere. There is also evidence that the areas covered by thunderstorm rainfall events may expand. The net effect is increasing volumes of water and multi-watershed flooding. Can our aging urban storm water infrastructure adequately collect and convey today's deluges? Can our reservoir management practice accommodate these deluges?

There will always be a worse flood. Instead of building fail-safe infrastructure, we need to consider safe-to-fail. For example, levees build resistance to flooding, but will be overtopped when the worse flood occurs. Letting some water flow though our infrastructure will incur some disruption, but also turns a catastrophic event into a managed event. Deep integration of science and practice around these 'graceful failure' solutions can improve protection of lives and property.

Answers to Questions Submitted by Ms. Kendra Horn

This past May, Oklahoma was hit by a series of tornadoes and flash flood events, which damaged thousands of homes and led to a State of Emergency being declared in every single one of our 77 counties. To my knowledge, the General Accountability Office has never undertaken a comprehensive study of the natural disaster trends facing our country, including the consequences to financial services, local economies, personal property and communities.

a. Considering the rise of weather events and natural disasters happening, do we need a comprehensive study and how valuable would it be to our understanding?

Today's infrastructure was built for yesterday's climate. How is our aging infrastructure combining with climate change and population change to driving disaster trends? While a number of efforts exploring disaster trends are underway, the effort is scattered across universities, research centers and other institutions. And many knowledge gaps remain. For example, we don't fully understand the process that causes people to permanently leave an area impacted by extreme weather. We also don't fully understand the response of our social and built infrastructure to the many different forms of extreme weather, nor can we best pinpoint the most effective adaptations. Answering these questions requires a convergence of scientific, engineering and emergency management expertise. *A* center for convergent science and practice where the diversity of required expertise and datasets come together has the potential to transform our understanding of trends in extreme weather impacts facing our nation.

In Oklahoma, many point to tornadoes as the largest cause for concern. However, the flash flooding we experienced back in May caught many families unaware and unprepared. We see rises in these scenarios not just in Oklahoma, but across the nation. The National Flood Insurance Program designates Special Flood Hazard Areas to illustrate flood risks, which homeowners are required to purchase flood insurance. However, these maps are often old and outdated. This can give many homeowners a false sense that properties outside these areas do not need flood insurance.

a. How can we enhance the maps and data we currently use?

Life threatening flash flooding and riverine flooding events are on the rise. There is mounting evidence that the heaviest rain rates will rise by about 4% for every degree Fahrenheit warming of our atmosphere. There is also evidence that summertime thunderstorms may become larger, and therefore produce greater volumes of rainfall in brief periods of time. We live in a new era of flood risk. Heavier deluges are combining with our rising populations and expanding urban centers, bringing new vulnerabilities to our homes and livelihoods.

Flood maps are key tools for the protection of life and property. But only if they accurately convey the risk. Maps of riverine and rainfall-driven (flash) flooding, however, are only partially complete nationwide, and can be based on oversimplified characterizations of how floods work. Accuracy is particularly poor for the smaller catchments. And these small catchments typically thread through urban areas. For some of these smaller catchments, the flood risk is missing entirely.

It's essential for the science to continue to empower communities. Recent advances in flood plain mapping are providing more comprehensive views of our flood risk. These new views find $5.5 trillion of assets lie within our 100-year flood plains. And we know that the 100-year flood has about a 25% likelihood of occurring within the typical 30-year mortgage. These improved maps show the total U.S. population exposed to serious flooding is 2.6-3.1 times higher than previous estimates, and that nearly 41million Americans live in the 100-year flood plain.

b. How can we improve communications with homeowners and give them a better understanding of the risks they may face?

Weather scientists and forecasters often ask 'Why don't people respond to the forecast the way I think they should?'. This is a problem of risk communication. Experts in risk communication at the National Center for Atmospheric Research (NCAR) and its partner universities are investigating what constitutes an effective delivery of a forecast. While risk communication is not my primary expertise, my understanding is that key requirements are that the forecast information is timely, relevant, salient, usable and comes from a trust worthy source.

Scientists and forecasters must work with communities to understand how to produce effective and equitable risk communication. One example from NCAR called Rising Voices brings Indigenous and other scientific professionals, tribal and community leaders, environmental and communication experts, students, educators, and artists from across the US and around the world, to assess critical community needs and communication and adaptation solutions. Thinking about flood risk communication in Oklahoma, physical and social scientists and forecasters must work with local communities to first understand local risk perceptions, priorities, and adaptive capacity of the at-risk communities. Only then can we begin to improve communication.

Many of the witnesses wrote in their testimony about the need for continued support for observational infrastructure to both provide real-time weather information and to assimilate into weather prediction models. Apart from observational technology, witnesses also wrote of the need to maintain high performing computing, data assimilation resources, and support through research grants.

a. What gaps exist in our observational infrastructure that need further investment?

One of the major challenges for forecast models is the representation of severe thunderstorms. This is particularly relevant for a thunderstorm-prone state like Oklahoma Severe thunderstorms organize across a range of scales from the large squall-line events that rake almost the entire state down to the localized 10-minute deluge. While our ability to forecast severe thunderstorms has improved substantially, detailed observations are needed to keep our forecasts on track in real-time. These observations are also required to improve our understanding of how severe thunderstorms work and how we can best represent them in our forecasts.

The Oklahoma Mesonet is a world leader in providing dense weather surveillance with quantifiable impacts on severe weather forecasting. *A* three-dimensional Mesonet scaled up across the nation has the potential to transform forecast accuracy through surveillance of the lowest couple of kilometers of the atmosphere. Adequate observation of the full atmosphere would require building Next Generation Phased array radars and funding for satellite-based hyperspectral sounding data. These in-situ surface, and remotely sensed radar and satellite data platforms would scan storms and their environments in unprecedented detail.

b. What are the most pressing needs in supercomputing for the Weather Enterprise?

Scientists are at the cusp of a revolution in their forecasting ability. This is being driven by advances in our understanding of how extreme weather works, but also by advances in supercomputing power. This power is about to enable us to produce weather forecasts in unprecedented detail. For example, forecasts will be able to provide information on straight-line winds pushing ahead of severe thunderstorms and the likelihood of localized intense rainfall cores. Not only that-this computational power will allow us to make large numbers of forecasts that map out a large range of possible scenarios. These scenarios, together with assigned confidence, has the potential to bring huge gains to the decisionmaking process of emergency responders.

But such detailed forecasts generate vast quantities of data. The data is so large we can't even see it all. A transformation is therefore needed in our ability to automate analysis and presentation of all this forecast data. Resources are critically needed to develop automated analysis techniques based on deep learning that can mine these mountains of data.

c. What role do advances in high performance computing and data assimilation capacity play in improving long-term forecasts, including sub-seasonal to seasonal forecasting?

Building on my response above, these detailed forecasts enabled by high-performance computing are only as good as the data fed into them. A goal of data assimilation is to provide the forecast model with a snapshot of the state of the atmosphere for the forecast model to advance forwards in time. Better snapshots lead to better forecasts. But only if the snapshots cover the key regions and phenomena.

Sub-seasonal to seasonal (S2S) forecasts are different to a typical weather forecast in that weather forecasts give the details of a weather event whereas S2S forecasts give likelihoods of a weather event. For example, a weather forecast may tell you that it will rain at 7pm on

Tuesday. An S2S forecast, on the other hand, may tell you that the chance of rain this month is higher than normal. The skill of S2S forecasts arise from persistence of weather patterns such as a stubborn region of high pressure, known cycles of atmospheric and oceanic patterns such as El Nino, and long-term trends. This means that for S2S forecasts, the snapshot must not just cover the local atmosphere but expand to see the global atmosphere and also the global oceans, and in some cases other components of the earth system such as land surface temperature and moisture and sea-ice cover. Creating the snapshots and ingesting them into the S2S forecast are incredibly computationally expensive. Advances are only possible through federally funded computational infrastructure.

Answer to the Question Submitted by Mr. Dan Lipinski

1. As Dr. Shepherd noted, climate change is no longer just about polar bears, but is now a "kitchen table" issue. During this hearing, we discussed many implications of climate change for coastal regions, but I represent an inland district in northern Illinois. A recent Assessment of the Impacts of Climate Change on the Great Lakes released by the Environmental Policy Law Center indicates that "Extreme weather events have already taken their toll on the MidWest" How can we expect extreme weather events to impact the Midwest/Great Lakes regions as the climate continues to change?

The Great Lakes region is no stranger to extreme weather. In addition to facing the extreme weather events that other regions face such as heatwaves and flooding rains the Great Lakes themselves can influence the local climate. The Lakes can cool the most intense heatwaves but may make them more humid. The Lakes can also warm the coldest winter air-masses, at least until lake ice cuts off the warmth.

Extreme weather can impact the region's drinking water systems, fisheries, recreation, and industry. Now enter climate change- a growing and pervasive threat multiplier. For the U.S. as a whole we have already detected a signal of climate change in the intensity of deadly heatwaves and flooding rains. Given that today's weather events operate within a warmer and more-moist atmosphere, it's inconceivable that the region's extreme weather has not changed. And further changes are expected.

Projected changes call for about 40 more days each year that exceed 90°F over Northern Illinois by mid 21st Century compared to the late 20th Century. These hot days will concentrate over our urban centers due to the storage and release of heat in our concrete infrastructure- the urban heat island effect.

Given the high likelihood that heavy rainfall has become even heavier in recent decades, design criteria based on historical data are already out of date. This introduces unknown vulnerability in this new era of extreme rainfall. The expected continuation of the changes we've already experienced may further limit the ability of today's water management systems to meet requirements.

Now consider severe thunderstorms. The peak of summer has become significantly more favorable for tornadoes over the Great Lakes region over the period 1979-2017. But we don't yet understand why. Further investigation is needed to understand the role of climate change, to understand whether this trend will persist in the future.

With warming, we expect fewer winter storms overall as more ice and snow events transition to rain events. However, there will still be events cold enough for ice and snow, and when these occur the air will be more humid. As a result, there is mounting evidence that we may experience more of the large, crippling snow storms over major urban centers.

The region must be able to maintain or even improve current levels of protection against extreme events in the face of population change and climate change. An in-depth investigation into the likely impacts from the collision of climate change and the region's population, infrastructure and industry is urgently needed.

Responses by Dr. Adam Sobel

Dr. Adam H. Sobel, Professor, Applied Physics & Applied Mathematics and Earth & Environmental Sciences, Columbia University, before the House, Science, Space, and Technology Committee, House of Representatives, Washington, D.C., September 26, 2019

Responses to Questions for the Record, October 29, 2019

Questions Submitted by Chairwoman Eddie Bernice Johnson

1) The EPIC program should do whatever it can to foster collaboration between the public, private and academic sectors. Good weather forecasts are an essential public good, and the weather, water and climate enterprise in the US must, as it has been in the past, be built on the foundation of public, government-funded observations, models, and infrastructure. That said, the private sector in our field is growing rapidly, creating new markets, goods and services, and academia has a great deal of talent and knowledge to contribute, not least in the form of energetic young students. A successful EPIC program will integrate all these capabilities. Academia should be brought in through a vigorous program of external funded research. In particular, I endorse my colleague Dr. Shepherd's recommendations to involve the American Meteorological Society's Commission on the Weather, Water and Climate Enterprise and the University Corporation for Atmospheric Research.
2) Although I am not aware of specific studies on this question, correlated or compound events are surely more challenging to forecast than single events are not necessarily because of any special difficulties (though there may be some), but, at least, simply because forecasting multiple things at the same time is

generally more difficult than forecasting one thing. If multiple events (or multiple aspects of one event) have to be forecast correctly, the chances of errors increase over those of a forecast of only one of them. Some research has already been done on some kinds of compound events, such as those involving heat and drought, or heat and humidity. For others almost none exists: for example, multiple severe tropical cyclones landfalling in the US in a single year (a challenging event for the insurance industry). In some kinds of events, the physical processes may not be fundamentally different from those involved in univariate events, while in others the compound or correlated nature poses unique research questions, such as: do events nearby in space and time influence each other, or are they simply both consequences of some larger-scale forcing? A program of federal funding specifically for compound and correlated events would, of course, allow the research community to address these gaps.

3) The most important thing is to maintain the existing network to insure the continuity, integrity, and consistency of long-term climate records, while also adding new capabilities as technology allows. The network of satellites, radiosonde and surface stations, radars etc. should be maintained with no gaps in coverage. A good recent study of the issue is here: https://agupubs.onlinelibrary.wiley.com/doi/full/l0.1002/2017EF000627.
4) High performance computing and data assimilation are both essential to forecasting and modeling of extreme weather. Forecasts of extremes are particularly sensitive to model resolution, for example, which is directly limited by computational resources. Investment in these areas is absolutely critical. My expertise, however, is not directly in these areas and so I do not have specific recommendations.
5) Climate change and extreme weather, together and separately, are most appropriately viewed as risk problems. We cannot ask for precise deterministic predictions, but rather only probabilistic estimates with uncertainties; decisions about disaster risk reduction

and climate adaptation should be made appropriately based on those. For probabilistic risk assessment of extreme events, the insurance industry uses what are known as "catastrophe models," but thus far those are both a) mostly proprietary and b) unable to handle climate change. In my written testimony (pp. 8-10), I advocated for federal support for the development of open-source catastrophe models which could be used by industry (not just insurance, but also climate-related financial disclosure and other applications), government, and academia. This is essential as the nation and the world move into a future of increasingly uncertain changes in extreme weather.

6) I support the recommendations of the recent National Academy study on seasonal-to-subseasonal prediction: https://www.nap.edu/catalog/21873/next-generation-earth-system-prediction-strategies-for-subseasonal-to-seasonal.

Question Submitted by Mr. James Baird

1) The event referred to by the Congressman, an entire season of high rainfall, is not a single weather event but rather a climate event with a seasonal time scale. Prediction of such events is precisely in the domain of "seasonal to subseasonal (S2S)" prediction. Forecasts on this time scale remain a frontier in the field and one where further investment could bring great benefits. As in my response to the last question from Chairwoman Johnson, I refer to the recent National Academy report, and endorse its recommendations on this subject: https://www.nap.edu/catalog/21873/next-generation-earth/system-prediction-strategies-for-subseasonal-to-seasonal.

Question Submitted by Mr. Michael Waltz

1) Florida could lead the way on climate adaptation and resilience by beginning an honest and serious discussion about the likelihood

that some areas of the state will, perhaps within a few decades, not be viable due to sea level rise and persistent flooding. In some areas, managed retreat may be necessary. In the shorter term, reform of the National Flood Insurance Program might lead to less development in the most vulnerable areas.

Questions Submitted by Ms. Kendra Horn

1) I believe that a comprehensive study of changing extreme weather and natural disaster trends by the General Accountability Office is an excellent idea and would be valuable for the nation.
2) a) NFIP flood maps in the coasts do not consider sea level rise, and I believe that maps inland do not account for the changing probabilities of extreme rainfall due to climate change. The latest climate science should be incorporated into these maps. New satellite methods can also be used to map inundation for recent or future events, increasing coverage and accuracy.
   b) To educate homeowners, social scientists, natural scientists, and media professionals could be engaged in a national program of risk communication around extreme weather and climate change.
3) These questions overlap with those asked by Chairwoman Johnson, please see my answers above.

Responses by Dr. Berrien Moore

House Committee on Science, Space, and Technology, "Understanding, Forecasting and Communicating Extreme Weather in a Changing Climate," Questions for the Record to: Dr. Berrien Moore, Director, National Weather Center, University of Oklahoma

Submitted by Chairwoman Eddie Bernice Johnson

1) How can the EPIC program foster collaboration between the public, academic, and private sectors of the weather enterprise?

What role can and should the academic and research community play in the development of the governance, goals, and implementation of the EPIC program?

2) How can improvements to subseasonal to seasonal (S2S) forecasting and prediction help us better prepare for future extreme weather events? What resources are needed to improve these long-term forecasts?

Questions by Chairwoman Eddie Bernice Johnson, Response to Question 1, Berrien Moore III

The EPIC program should offer a unique opportunity for collaboration in advancing not only the "dynamic core (FE-3)," which expresses mathematically the underlying physics of the nation's next generation numerical weather prediction models, but also, improving the algorithms that allow Incorporation (data assimilation) of observations into this model and an opportunity to develop a coherent cost-effective plan for enhancing the observational base. This will be a unique opportunity for public, academic, and private sectors to work as a community in the research effort as well as charting and realizing the pathway to operations; however, this will require the supporting infrastructure that would allow this unique collaboration, which will need to take place in the cloud. This has never really been done before, but we should not let that fact stall the enterprise; we must avoid paralysis by analyses; the perfect must not be allowed to be the enemy of the good; in other words, we need to get “cracking.”

The governance, goals and implementation of the EPIC program requires active engagement of the academic and research community. This has to be achieved by a combined bottom-up approach, which incorporates the thoughts of the academic and research communities as well as a "top-down" approach that is developed from a coordinated engagement of the leadership within NOAA and the academic communities (and public as appropriate). The August EPIC Workshop in Boulder collected, I believe, what is needed *from* the "bottom-up" approach; we now need to engage the leadership. One avenue would be to reach out directly to the leadership in relevant NOAA Cooperative Institutes. However, time is going by and we need, again, to get "cracking.

Questions by Chairwoman Eddie Bernice Johnson, Response to Question 2, Berrien Moore III

In thinking about the value of subseasonal to seasonal (S2S) forecasts, the landscape is extensive from prediction of vector-borne disease outbreaks to how much salt the Mayor of Philadelphia such buy for the winter. In fact, the space of applications is so large one can become lost; therefore, let me constrain the space to agriculture (which is large). Stating it simply:

With improved S2S forecasts of:

- Temperature
- Precipitation
- Wind speed
- Relative humidity
- Soil temperature
- Projected dates of first/last freeze
- Probability of disruptive events -flood, drought, heat waves, freeze

Farmers can better understand:

- What crops and varieties to plant
- When to apply irrigation, nutrients, pesticides & herbicides
- Projected crop yields to support decisions about food production and distribution

Looking to a Congressional Briefing on the value of subseasonal to seasonal forecasts in April 2018,[90] the value chain was extensive and parts highly important to Oklahoma and the mid-west generally. It was noted at

[90] https://news.ucar.edu/132389/ucar-congressional-briefing-subseasonal-seasonal-forecasts.

the Briefing that Chad McNutt, principal and co-founder of Livcstock Wx, had recently stated that:

> "Agriculture sectors like the cattle industry need sustained support for research into improved subseasonal to seasonal forecasting... The forecasts have real economic implications for producers."

Looking at the closely related field of water management in the context of climate change, we find papers and statements such as:

> "Seasonal streamflow predictions provide a critical management tool for water managers in the American Southwest. In recent decades, persistent prediction errors for spring and summer runoff volumes have been observed in a number of watersheds in the American Southwest. While mostly driven by decadal precipitation trends, these errors also relate to the influence of increasing temperature on streamflow in these basins. Here we show that incorporating seasonal temperature forecasts from operational global climate prediction models into streamflow forecasting models adds prediction skill for watersheds in the headwaters of the Colorado and Rio Grande River basins. Current dynamical seasonal temperature forecasts now show sufficient skill to reduce streamflow forecast errors in snowmelt-driven regions. Such predictions can increase the resilience of streamflow forecasting and water management systems in the face of continuing warming as well as decadal-scale temperature variability and thus help to mitigate the impacts of climate nonstationarity on streamflow predictability."[91]

In a recent report of the National Academies of Science, which presented strategies for how to move towards the next generation Earth system prediction models that can provide more accurate predictions for the subseasonal to seasonal time-scales, concluded that in a decade from now Intermediate range (S2S) forecasts are expected to be as widely used as weather forecasts today.

---

[91] https://agupubs.onlinelibrary.wiley.cornldoi/pdfll 0.1002/2017GL076043.

Meeting the challenge of increasing S2S Forecast Skill, the "subchallenges" can be divided as follows:

- Improving Earth System Models:
    - Critical Earth system processes need to be parameterized in models - represented through a set of equations rather than being resolved -for the foreseeable future
    - Improving physical parameterizations is fundamental to increasing S2S forecast skill
    - Key Strategy. Improve parameterization of unresolved (e.g., subgrid scale) processes, both within S2S system submodels, and holistically across models to better represent coupling in the Earth system. This also will require significant interaction with expertise in Computer Science as well as Cloud Computing and Machine Learning
- Data Assimilation: the process of initializing and updating Earth system models to be consistent with observations
    - Capturing the interactions among components of the Earth system – a process known as "coupling" – is central to strengthening S2S predictions
    - Developing data assimilation methods that allow for better coupling is important to improving S2S forecast systems
    - Examine the value of model diversity, as well as the impact of various selections and combinations of model resolution, number of ensemble perturbations, length of lead, averaging period, length of retrospective forecasts, and options for coupled sub-models.
    - Key Strategy. Advance the development of strongly-coupled data assimilation and quantify the impact of such advances on operational S2S forecast systems
- Extreme Events:
    - The prediction of extreme events is often more meaningful to stakeholders than predicting mean conditions.

- o Forecasting the probabilities of such events on S2S timescales will likely involve identifying windows in time when the predictability of such events is elevated.
- o Key Strategy. Focus predictability studies, process exploration, model development and forecast skill advancements on high impact S2S "forecasts of opportunity" that in particular target disruptive and extreme events.

But not all of the challenges are mathematical and modelling. There are very real challenges on the social sidc. For instance, we need to

- Develop a body of social science research that leads to more comprehensive and systematic understanding of the use and barriers to use of seasonal and subseasonal Earth system predictions;
- Establish an ongoing and iterative process in which stakeholders, social and behavioral scientists, and physical scientists co-design S2S forecast products, verification metrics, and decision-making tools.

House Committee on Science, Space, and Technology, "Understanding. Forecasting and Communicating Extreme Weather in a Changing Climate," Questions for the Record to: Dr. Berrien Moore, Director, National Weather Center, University of Oklahoma

Submited by Mr. Michael Waltz

1) Florida is leading the country in resiliency and mitigation in many ways, including the wind standard in the state building code. How can Florida and the United States by more resilient to flooding and how should we rebuild infrastructure more resiliently?

Questions by Congressman Michael Waltz, Response to Questions, Berrien Moore III

I am not expert on flooding and resilient building matters; however, I would think that there are several "First Order" issues that we best have right:

1) Are flood zone maps up-to-date with respect to decadal precipitation patterns?
2) Do they include the impact of the "built environment"?
3) Do they include sea-level rise and predications of future sea-level heights?
4) Is storm surge accurately represented and is the built environment considered.
5) Is subsidence included in storm-surge predictions and flooding.
6) What is the impact of upstream flood control on down-stream communities?

House Committee on Science, Space, and Technology, "Understanding, Forecasting, and Communicating Extreme Weather in a Changing Climate," Questions for the Record to: Dr. Berrien Moore, Director, National Weather Center, University of Oklahoma, Submitted by Mr. James Baird

1) In my home state of Indiana, we received nearly 19 inches of rain this spring from March 20-June 21st. That's approximately five inches more than last year. Many in the rural fanning communities I represent struggled to get their crops in the ground and faced a very uncertain and challenging planting season. As a fanner myself, I understand that struggle. Which brings me to my question:

Can you elaborate on how atmospheric modeling is currently used in extreme weather forecasting? What kind of additional information is needed to improve atmospheric modeling? And what research priorities would you ask Congress to support?

Questions by Congressman James Beard, Response to Questions, Berrien Moore III

We use the same Numerical Weather Prediction (NOW) models for extreme weather precipitation as with standard NWP modelling; however, your questions does cause me to suggest some paths that might help in extreme rainfall predictions and these tie to my responses on Subseasonal to Seasonal weather prediction. Namely, focus predictability studies, process exploration, model development and forecast skill advancements on high impact extreme events.

I think the single most Important need to Improve atmospheric modeling Is to fly an advanced hyperspectral sounding instrument in geostationary orbit (between GOES 16 and GOES 17. Only an advanced hyperspectral sounding can provide the needed detailed moisture and dynamical motion Information with high temporal and high spatial resolution into the boundary layer. Advanced IR sounders, such as the Atmospheric Infrared Sounder (AIRS) on Aqua and the Cross-track Infrared Sounder (CriS) on Suomi-NPP (SNPP) and Joint Polar Satellite System (JPSS) series, do provide high vertical resolution atmospheric sounding information that are already improving the forecast skill in numerical weather prediction models. However, those high-quality measurements are only available in sun-synchronous LEO at a revisit rate of every 8 hours. For ground-breaking severe weather and near-term forecasting of threatening weather in the US, near real-time soundings are necessary to detect changes in the atmosphere as *they occur,* which requires a sounder in geostationary orbit over CONUS.

Regarding Research Priorities:

- As you know well, the Weather Research and Forecasting innovation Act (https://www.congress.gov/biU/115th-congresslhouse-bill/353) requires NOAA's Office of Oceanic and Atmospheric Research (OAR) to conduct a program aimed at developing *"an improved understanding of forecast capabilities for atmospheric events and their impacts, with priority given to the development of more accurate, timely, and effective warnings and forecasts of high impact weather events that endanger life and property."* In January 2019, congress passed a related bill that authorizes an "Earth Prediction Innovation Center (EPIC)" focused on "advancing weather modeling skill, reclaiming and maintaining international leadership. In the area of numerical weather prediction, and improving the transition of research into operations." For severe short term weather, EPIC must be supported and must be realized (See my answers to Chairwoman Johnson's Question Number 1).

- The H.R.353 bill also asks for the NWS to *"collect and utilize information to make reliable and timely foundational forecasts of subseasonal and seasonal temperature and precipitation. Subseasonal forecasting is forecasting weather between two weeks and three months and seasonal forecasting is between three months and two years.* Predictions at these time scales are critical for many practical applications, such as e.g., water management, prevention of vector-borne diseases like west Nile virus, energy management, and agriculture. We need a similar EPIC-Like thrust for subseasonal seasonal to seasonal forecasting of temperature and precipitation.

House Committee on Science, Space, and Technology, "Understanding, Forecasting, and Communicating Extreme Weather in a Changing Climate," Questions for the Record to: Dr. Berrien Moore, Director, National Weather Center, University of Oklahoma

Submitted by Ms. Kendra Hom

1) This past May, Oklahoma was hit by a series of tornadoes and t1asb flood events, which damaged thousands of homes and led to a State of Emergency being declared in every single one of our 77

counties. To my knowledge, the General Accountability Office has never undertaken a comprehensive study of the natural disaster trends facing our country, including the consequences to financial services, local economies, personal property and communities.

   a. Considering the rise of weather events and natural disasters happening, do we need a comprehensive study and how valuable would it be to our understanding?

2) In Oklahoma, many point to tornadoes as the largest cause of concern. However, the flash flooding we experienced back in May caught many families unaware and unprepared. We see rises of these scenarios not just in Oklahoma, but across the nation. The National Flood Insurance Program designates Special Flood Hazard Areas to illustrate flood risks, which homeowners are required to purchase flood insurance. However, these maps are often old and outdated. This can give many homeowners a false sense that properties outside these areas do not need flood insurance.
   a. How can we can enhance the maps and data we currently use?
   b. How can we improve communications with homeowners and give them a better understanding of the risks they may face?
3) Many of the witnesses wrote in their testimony about the need for continued support for observational infrastructure to both provide real-time weather information and to assimilate into weather prediction models. Apart from observational technology, witnesses also wrote of the need to maintain high performing computing, data assimilation resources, and support through research grants.
   a. What gaps exist in our observational infrastructure that need further investment?
   b. What are the most pressing needs in supercomputing for the Weather Enterprise?
   c. What role do advances in high performance computing and data assimilation capacity play in improving long-term forecasts, including sub-seasonal to seasonal forecasting?

Questions by Congresswoman Kendra Horn, Response to Questions, Berrien Moore III

GAO Study. I think it would be very valuable to have a comprehensive study on natural disaster trends. In fact, this, like to census, should become a "Decadal Requirement."

Flood Maps

I think my answer to Congressman Waltz is relevant. There are several "First Order" issues that we best have right:

1) Are flood zone maps up-to-date with respect to decadal precipitation patterns?
2) Do they include the impact of the "built environment"?
3) Do they include sea-level rise and predications of future sea-level heights?
4) Is storm surge accurately represented and is the built environment considered.
5) Is subsidence included in storm-surge predictions and flooding.
6) What is the impact of upstream flood control on down-stream communities?

Improve Communications

Frankly, I am not certain, but perhaps using models and data would could create an "on-line" data base into which one would only need to submit an address, and the response would be “flood forecast" under different climate and precipitation and storm- surge forecasts-in other words a dynamic model/data derived interactive database.

Gaps in Observational Infrastructure

See my testimony on next generation weather radars and a hyperspectral sounder in geostationary orbit.

Supercomputing

I think the gaps is not so much in "big computer iron" but rather in the infrastructure supporting Cloud Computing. NOAA's Unified Forecast

System is an important start, but it needs to be enhanced and made availability. There is also the outstanding issue of costs, which have not been adequately addressed.

What Roles ...

See my response to Chairwoman Johnson's Question #2.

Responses by Dr. Ann Bostrom

House Committee on Science, Space, and Technology, "Understanding, Forecasting, and Communicating Extreme Weather in a Changing Climate," Questions for the Record to: Dr. Ann Bostrom, Weyerhaeuser Endowed Professor in Environmental Policy, University of Washington

Submitted by Chairwoman Eddie Bernice Johnson

1) How can the EPIC program foster collaboration between the public, academic, and private sectors of the weather enterprise? What role can and should the academic and research community play in the development of the governance, goals, and implementation of the EPIC program?

There appears to be some degree of consensus in science advisory groups that the EPIC program will foster U.S. leadership in weather modeling if it successfully engages researchers from across academia, government and other research organizations in a concerted and focused effort. The academic and broader research community should partner with NOAA, NASA, USGS and other federally engaged entities to develop the governance, goals and implementation of the EPIC program. For this to happen successfully, it will require acknowledging the differences in the conditions and incentives that govern researchers across sectors; inviting, incentivizing, and respecting diverse contributions; and assuring rigorous, community-vetted processes for developing, testing and validating models with high quality data.

- Include widely respected academic scholars in the leadership and development of EPIC, and if possible leaders who have worked across academia, government, nonprofits, and/or industry.
- Develop training programs that support and integrate graduate student and postdoctoral research and training with a diverse group universities.[92]
- Guide the implementation of EPIC with the best available evidence and science on interdisciplinary and cross-sectoral innovation and teamwork.[93]

Although there has been mention of "no bricks and mortar" for EPIC, co-location and face-to- face meetings can be invaluable in facilitating innovation and training, and incorporation of existing NOAA Collaborative Institutes, or other organizations such as UCARINCAR, may afford opportunities to identify EPIC facilities within one or more of these. As important are the virtual coordination mechanisms and support for developing an active, inclusive EPIC community. One element of this is making EPIC code open source. The LEGEND Act (S.2597) would require

---

[92] The National Academies have published numerous reports on Team Science and on innovation processes. See, e.g., National Academies of Sciences, Engineering, and Medicine. 2016. Next Generation Earth System Prediction: Strategies for Subseasonal to Seasonal Forecasts. Washington, DC: The National Academies Press. https://doi.org/10.17226121873. On page 236 the report notes: "Modeling centers outside of the United States, such as the European Centre for Medium-Range Weather Forecasts (ECMWF), have attempted to attract and retain more people in [subseasonal to seasonal] model development work by appointing model developers to 5-year terms, which is longer than typical research grant cycles in the United States (3 years). ECMWF offers strong incentives to attract top scientists to model development, such as access to excellent facilities, excellent tools (e.g., what some regard as the most advanced numerical weather prediction model in the world), and high, tax-free salaries. Furthermore, the inclusion of highly reputed scientists within the limited staff (150 staff members and 80 consultants) encourages a stimulating environment where delivering end-use forecasting products and conducting cutting-edge scientific research are valued and directly coupled."

[93] E.g., National Academies of Sciences, Engineering, and Medicine 2019. Adapting to the 21st Century Innovation Environment: Proceedings of a Workshop in Brief. Washington, DC: The National Academies Press. https://doi.orgll 0.17226/25384.

models developed by the EPIC program to be open source. Another critical element is to provide sufficient resources to fund the enormously increased computational resources required for EPIC to be successful.

For additional comments on this topic, please see the recent letter report and presentation by the Environmental Information Services Working Group on EPIC to the NOAA Science Advisory Board, available on the NOAA SAB website:

EISWG Letter Report on EPIC to the SAB

SAB September 9th Presentation by the EISWG Co-chairs

2) What are the most pressing high performance computing and data assimilation needs within the weather enterprise? How critical are these resources to the improvement of our extreme weather forecasting capabilities, as well as the communication of these forecasts?

High performance computing and data assimilation needs are well beyond my expertise, but are referenced in the NASEM report cited above 1 as well as in the EISWG letter report and SAB presentation referenced in the previous question.

There may be some implications from communications needs for high performance computing, due to increased demand for dynamic, timely visualizations of forecasts. Recent reports have called for more research on geotargeting warnings (NASEM, 2018),[94] which suggests a need for more research on geotargeting the forecasts that give rise to those warnings, especially with the advent of warn-on-forecast. Widescale, rapid, visual communication of geographically and temporally microtargeted forecasts (as was illustrated in the Dallas-Fort Worth Collaborative Adaptive Sensing of the Atmosphere (CASA) Urban Test Bed) is likely to require

[94] National Academies of Sciences, Engineering, and Medicine. 2018. Emergency Alert and Warning Systems: Current Knowledge and Future Research Directions. Washington, DC: The National Academies Press. https:lldoi.org/10.17226124935.

highspeed data assimilation, significant computational resources, and considerable communications bandwidth.

3) What do we know and what gaps remain in knowledge about disproportionate impacts of extreme weather on vulnerable populations in the U.S.?

There is considerable evidence that extreme weather disproportionately affects vulnerable populations, in the U.S. as elsewhere.[95] For example, children are particularly vulnerable,[96] and their schooling can be severely disrupted by extreme weather events.

Many policies are designed to help the vulnerable out after disasters.[97] However, recent evidence suggests these policies in fact may inadvertently be disproportionately helping the wealthy, increasing the gap between rich and poor.[98]

For example, in a comprehensive study published earlier this month, compelling evidence is presented that richer counties are more likely to implement buyout counties, but the areas within the counties where buyouts occur are poorer, less educated, and more racially diverse.[99]

4) In what ways might social and behavioral sciences be strengthened to advance understanding, forecasting, and communicating

[95] See page 42 in National Academies of Sciences, Engineering, and Medicine (2018). *Integrating Social and Behavioral Sciences within the Weather Enterprise.* Washington, DC: The National Academies Press https://doi.org/10.17226/24865.

[96] https://www.phe.gov/Preparedness/legallboards/naccd/meetings/Documents/naccd-hs-wg-rpt 082017.pdf.

[97] Federal policy environment influencing disaster recovery (Appendix A of this report: Institute of Medicine. 2015. Healthy, Resilient, and Sustainable Communities after Disasters: Strategies, Opportunities, and Planning for Recovery. Washington, DC: The National Academies Press. https://doi.org/10.17226/18996.)

[98] Howell, J., & Elliott, J. R. (2019). Damages done: The longitudinal impacts of natural hazards on wealth inequality in the United States. *Social Problems, 66(3),* 448-467.

[99] Mach, K. J., Kraan, C. M., Hino, M., Siders, A. R., Johnston, E. M., & Field, C. B. (2019). Managed retreat through voluntary buyouts of flood-prone properties. Science Advances, 5(10), eaax8995. https://advances.sciencemag.org/content/5/1 O/eaax8995 See also: https://www.npr.org/2019/10/09/767920427/sweeping-study-raises-questions-about-who-benefits-from-buyouts-of-flood-prone-h.

extreme weather in a changing climate, and how can federal resources help support this?

As summarized in *Integrating Social and Behavioral Sciences within the Weather Enterprise,*[100] efforts are needed-especially by NOAA-both to build capacity, and to advance what we know on critical topics.

To paraphrase and expand on the report's recommendations with regard to building capacity:

A planning process should be initiated immediately by Federal agencies and private-sector weather (and climate) companies together with leading social and behavioral scientists with diverse expertise, to identify specific investments and activities that collectively advance research at the social and behavioral science-weather and climate interface. Beyond direct training and research funding, activities that support capacity building should be considered in this planning process as well, including: assessing existing research efforts, setting research agendas, building community, improving information sharing, and developing methods to track funding support of social and behavioral weather and climate research activities.

It behooves NOAA to increase its efforts to build more sustainable institutional capacity for research and operations at the social and behavioral sciences (SBS)-weather-climate interface and to advance cooperative planning to expand SBS research among other federal agencies that play critical roles in weather- and climate-related research operations. In particular, this should include (but not be limited to) leadership from:

- NSF for a strong standing program that supports interdisciplinary research at the SBS-weather interface,
- FHWA for research related to weather impacts on driver choices and behaviors, and

[100] See pages 7-9 for a summary of recommendations on this topic from: National Academies of Sciences, Engineering, and Medicine (2018). *Integrating Social and Behavioral Sciences within the Weather Enterprise.* Washington, DC: The National Academies Press. https://doi.org/1 0.17226/24865. Hereafter NASEM 2018.

- FEMA for research on the social and human factors that affect weather readiness, including decisions and actions by individuals, communities, and emergency management to prepare for, prevent, respond to, mitigate, and recover from weather hazards.

Across all sectors, all parties in the weather enterprise should continue to develop and implement training programs for current and next generation workforces in order to expand capacity for SBS-weather-climate research and applications in the weather enterprise.

Specific critical research gaps identified in the report include (NASEM 2018, from the report summary):

- Research focusing on the how the weather enterprise system itself works, including "system-level studies of weather information production, dissemination, and evaluation; studies of how forecasters, broadcast media, emergency and transportation managers, and private weather companies create information and interact and communicate among themselves; studies of forecaster decision making, such as what observational platforms and numerical weather prediction guidance forecasters use and how they use them; studies of how to assess the economic value of weather services; and studies of team performance and organizational behavior within weather forecast offices and other parts of the weather enterprise."
- Research on how people assess and respond to the risks from extreme weather and climate change. There is a need for research on how to better reach and inform special-interest populations with unique needs, such as vehicle drivers and others vulnerable to hazardous weather due to their location, resources, and capabilities. More research is also needed on how specific social or physical contexts, prior experiences, cultural backgrounds, and personal values influence people's interest in, access to, and interpretation of weather and climate information, as well as their decisions and actions in response.

- Message design, delivery, interpretation and use. "Persistent challenges include understanding how communicating forecast uncertainties in different formats influences understanding and action; how to balance consistency in messaging with needs for flexibility to suit different geographical, cultural, and use contexts, including warning specificity and impact-based warnings; and how new communication and information technologies- including the proliferation of different sources, content, and channels of weather information-interact with message design and are changing people's weather information access, interpretations, preparedness, and response." (NASEM, 2018, pages 9-10). As technologies evolve, the communication potentials have also evolved, creating a dynamic communications environment whose characteristics SBS-research has yet to fully understand, although SBS-research can predict some aspects of its performance.[101]

Submitted by Ms. Kendra Horn

1. This past May, Oklahoma was hit by a series of tornadoes and flash flood events, which damaged thousands of homes and led to a State of Emergency being declared in every single one of our 77 counties. To my knowledge, the General Accountability Office has never undertaken a comprehensive study of the natural disaster trends facing our country, including the consequences to financial services, local economies, personal property and communities.
   a. Considering the rise of weather events and natural disasters happening, do we need a comprehensive study and how valuable would it be to our understanding?

GAO has published several reports in the last year concerning hazards and disasters, many related to extreme weather and climate events.

---

[101] National Academies of Sciences, Engineering, and Medicine.2017. *Communicating Science Effectively: A Research Agenda.* Washington, DC:The National Academies Press. https://doi.org/10.17226/23674.

However, we are far from having a comprehensive picture of the costs of disasters attributable to extreme weather and climate change.

There are multiple sources for data on disasters in the U.S., as you may be aware. Arizona State University now hosts the Spatial Hazard Events and Losses Database for the United States (SHELDUS, at https://cemhs.asu.edu/sheldus), which was developed by an Susan Cutter at University of South Carolina with National Science Foundation and USC funding and is now maintained by ASU Center for Emergency Management and Homeland Security. It includes county-level direct hazard loss data and insured crop loss data from 1960 to the present.

In another example, NOAA publishes information on billion dollar weather and climate disasters (see https://www.ncdc.noaa.gov/billions/mapping for a map view, and for time series see https://www.ncdc.noaa.gov/billions/time-series). NOAA analyses show that such costs are rising over recent decades. However, the authors note that they rely a compilation of data from numerous sources, for example information from state governments, not all of which is complete.[102]

Further, there are gaps in our understanding of impacts that scholars attribute to a lack of spatially explicit loss data.[103] One illustration of the potential consequences of having inadequate data is a lack of awareness of how the U.S. is performing relative to other countries. These and other analyses by Bakkensen and Mendelsohn[12] show that the U.S. stands out as an outlier, with disproportionately high damages from tropical cyclones in comparison to other developed countries.[104]

A study by GAO or NOAA of the sort suggested in the question might be helpful, to bring attention to such data gaps. Perhaps even more helpful initially would be an assessment of the quality and accessibility of temporally and spatially specific data on extreme weather and climate change impacts on individuals and communities in the U.S., and their adequacy for assessing trends, along with commitment to address any

---

[102] https://www.ncdc.noaa.gov/monitoring-content/billions/docs/lott-and-ross-2006.pdf.

[103] Bakkensen, L.A. and Mendelsohn, R.O., 2016. Risk and Adaptation: Evidence from Global Hurricane Damages and Fatalities. Journal of the Association of Environmental and Resource Economists 3(3): 555-587.

[104] See also my written testimony.

notable gaps that might slow adaptation and risk mitigation efforts. International efforts have been underway for some time to increase the quality and consistency of disaster loss data, but there is still a considerable amount of work to be done, including reaching agreement on concepts such as vulnerability and on how to measure them.

2. In Oklahoma, many point to tornadoes as the largest cause of concern. However, the flash flooding we experienced back in May caught many families unaware and unprepared. We see rises of these scenarios not just in Oklahoma, but across the nation. The National Flood Insurance Program designates Special Flood Hazard Areas to illustrate flood risks, which homeowners are required to purchase flood insurance. However, these maps are often old and outdated. This can give many homeowners a false sense that properties outside these areas do not need flood insurance.
   a. How can we can enhance the maps and data we currently use?

The lack of accurate mapping is a barrier to an effective insurance program.[105] Recent remapping efforts by FEMA have been contentious, and, some argue, subject not only to limitations stemming from lack of data and resources, but also subject to political pressures, biased toward more wealthy mortgage owners (who can afford appeals), and limited by their historical (rather than prospective) basis.[106] More accurate maps could be produced, for example, with Lidar and available or emerging regional climate trend information. Recent research looks at how better to integrate what we know about climate change and sea level rise into flood maps.[107] Putting more resources into such efforts and integrating their results into

---

[105] Kunreuther, H. (2018). Reauthorizing the National Flood Insurance Program. Issues in Science and Technology, 34(3), 37-50.

[106] Pralle, S. (2019). Drawing lines: FEMA and the politics of mapping flood zones. Climatic change, 152(2), 227-237.

[107] E.g., Amante, C. J. (2019). Uncertain seas: probabilistic modeling of future coastal flood zones. International Journal of Geographical Information Science, 33(11), 2188-2217.

flood risk mapping would likely reduce damages and costs in the longer run.

b. How can we improve communications with homeowners and give them a better understanding of the risks they may face?

Communications can be improved in numerous ways, some of which include making the communications more accessible and available, for example by matching the time scales (e.g., cumulative risk of flooding) with the decisions homeowners face (mortgage duration), or reaching out to them through communication channels they use most frequently. Other approaches might take advantage of changing the way the decision is framed, for example by changing the default decision so that people are more likely to opt into a more protective decision, such as buying insurance. We have learned a lot about specific elements of message design, such as the effects of different formats for communicating probabilities, but much more empirical research-and a more comprehensive and systematic approach to science and risk communication research-is needed.[108] Specific communications should be evaluated before they are widely adopted.[109]

3. Many of the witnesses wrote in their testimony about the need for continued support for observational infrastructure to both provide

---

[108] Spiegelhalter, D. (2017). Risk and uncertainty communication. Annual Review of Statistics and Its Application, 4, 31-60.

Also see pages 43-44 in: National Academies of Sciences, Engineering, and Medicine (2018). *Integrating Social and Behavioral Sciences within the Weather Enterprise.* Washington, DC: The National Academies Press. https://doi.org/10.17226/24865.

See page 29 in National Academies of Sciences, Engineering, and Medicine. 2017. *Communicating Science Effectively: A Research Agenda.* Washington, DC: The National Academies Press. https:/ldoi.org/10.17226123674.

And see pages 258-259 in: Bostrom, A., Bohm, G., O'Connor, R. E., Communicating Risks: Principles and Challenges, (2018). Chapter II in Eva Lermer, Martina Raue and Bernhard Streicher (Ed.), Psychological Aspects of Risk and Risk Analysis: Theory, Models, and Applications. Springer.

[109] Fischhoff, Baruch, Noel T. Brewer & Julie S. Downs, editors. (20 II). Communicating Risks and Benefits: An Evidence-Based User's Guide. U.S. Dept of Health and Human Services, Food and Drug Administration. https://www.fda.gov/ahout-fdalreports/communicating-risks-and-benefits-evidence-based-users-guide.

real-time weather information and to assimilate into weather prediction models. Apart from observational technology, witnesses also wrote of the need to maintain high performing computing, data assimilation resources, and support through research grants.

a. What gaps exist in our observational infrastructure that need further investment?

This is not in my area of expertise, but there are still radar shadows and other gaps in local data in some places, and there is a lack of mesolevel extreme weather observational infrastructure (e.g., other states do not have the mesonet Oklahoma has), such that not all areas of the country have equally good observational data. More generally, we still lack an understanding of cloud physics and other atmospheric phenomena for which continued advances in basic science and new measurement approaches to improve observational data are needed.

b. What are the most pressing needs in supercomputing for the Weather Enterprise?

This is outside my area of expertise, but as a member of science advisory panels for the Mesoscale & Microscale Meteorology Laboratory at the National Center for Atmospheric Research (NCAR) and for Environmental Research and Education at the National Science Foundation I've seen convincing presentations on the extraordinary computing resources required to run high resolution simulations and ensemble forecasts, and the importance of long-range planning to optimize how software interacts with hardware. While cloud computing has been envisioned for EPIC, NOAA relies on having advanced supercomputing resources for operational purposes. As has been suggested by NOAA leaders, it may well be to the nation's advantage to increase the compatibility between research and operational computing infrastructure (e.g., between NCAR and NOAA), in order to speed up advances in forecasting.

c. What role do advances in high performance computing and data assimilation capacity play in improving long-term forecasts, including sub-seasonal to seasonal forecasting?

This is also far outside my area of expertise. However, as indicated in our EISWG Letter Report on EPIC to the SAB, both appear essential to such improvements.

Appendix II: Additional Material for the Record

Report Submitted by Representative Kendra Horn
2018 Natural Hazard Report

Executive Summary

2018 was an eventful year worldwide. Wildfires scorched the West Coast of the United States; Hurricanes Michael and Florence battered the Gulf and East Coast. Typhoons and cyclones alike devastated the Philippines, Hong Kong, Japan and Oman. Earthquakes caused mass casualties in Indonesia, business interruption in Japan and structure damage in Alaska. Volcanoes made the news in Hawaii, expanding the island's terrain.

1,000-year flood events (or floods that are said statistically to have a 1 in 1,000 chance of occurring) took place in Maryland, North Carolina, South Carolina, Texas and Wisconsin once again. Severe convective storms pelted Dallas, Texas, and Colorado Springs, Colorado, with large hail while a rash of tornado outbreaks, spawning 82 tornadoes in total, occurred from Western Louisiana and Arkansas all the way down to Southern Florida and up to Western Virginia.

According to the National Oceanic and Atmospheric Administration (NOAA),[110] there were 11 weather and climate disaster events with losses exceeding $1 billion in the U.S. Although last year's count of billion-dollar events is a decrease from the previous year, both 2017 and 2018 have tracked far above the 1980-2017 annual average of $6 billion events.

In this report, CoreLogic® takes stock of the 2018 events to protect homeowners and businesses from the financial devastation that often follows catastrophe. No one can stop a hurricane in its tracks or steady the ground from an earthquake, but with more information and an understanding of the risk, recovery can be accelerated and resiliency can be attained.

This assessment covers an analysis of what the risk and exposure looked like, what happened during the event and the residential and commercial losses which occurred in the aftermath for each notable climatological event.

All the data in this report is current to November 30, 2018, unless denoted otherwise.

Hurricane (Tropical Cyclone)

The 2018 Atlantic hurricane season saw above average storm activity, deviating from earlier projections[111] at the start of hurricane season. The season ended with 15 named storms, eight of which were named hurricanes. Two of these, Hurricanes[112] Florence and Michael, made landfall along the U.S. This made 2018 the third consecutive season of above-average hurricane activity in the Atlantic.

Atlantic Hurricanes

At its most powerful, Hurricane Florence was a Category 4 storm. It eventually made landfall as a strong yet slow-moving Category 1 hurricane off the coast of North Carolina on September 14 with wind gusts up to 105

---

[110] https://www.ncdc.noaa.gov/billions/events.
[111] https://www.cpc.ncep.noaa.gov/products/outlooks/hurricane2018/May/hurricane.shtml.
[112] http://www.aoml.noaa.gov/hrd/tcfaq/A1.html.

mph. It caused significant damage in North Carolina, South Carolina and Virginia.

Table 1. Hurricane Florence residential and commercial loss

Peril	Residential and Commercial Total Losses in Billions (U.S. Dollars)
Wind	1 – 1.5
Flood	19 – 28.5
Total	20 – 30

Source: CoreLogic.

In these states, approximately 700,000 residential and commercial properties experienced catastrophic flooding and wind damage.

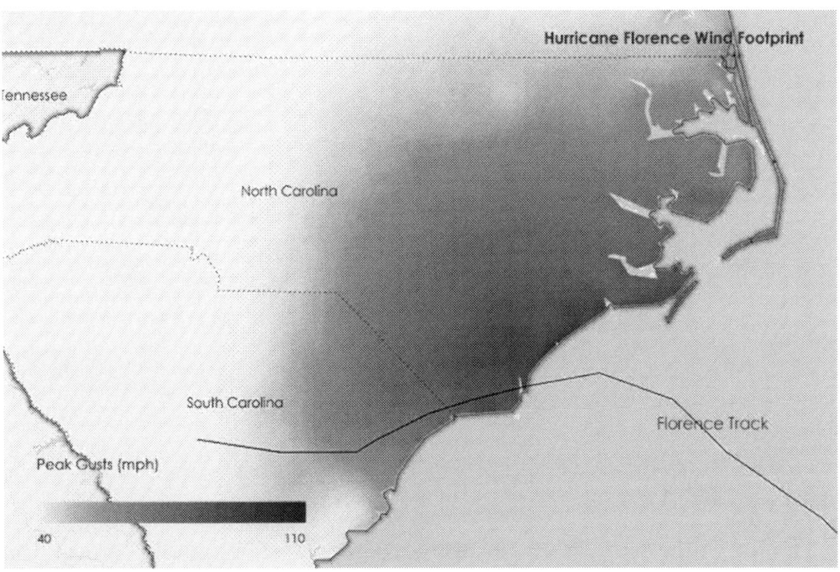

Source: CoreLogic, NHC.

Figure 8. Hurricane Florence Wind Footprint.

The largest losses were a result of inland flooding as Florence eventually stalled and moved slowly inland, causing a downpour on already water-logged land. Roughly 85 percent of residential losses from flooding were uninsured. Unlike wind damage, flood is covered separately from a standard homeowners' policy and is not mandatory to purchase outside of Special Flood Hazard Areas (SFHAs).

Weeks later, Hurricane Michael made landfall as a Category 4 hurricane along the Florida Panhandle on October 10. The maximum sustained wind speed at landfall was 155 mph, just 2 mph short of Category 5 classification. Michael is the strongest hurricane to make landfall in the Florida Panhandle since 1900 and the strongest hurricane to make landfall in the U.S. since Hurricane Andrew in 1992.

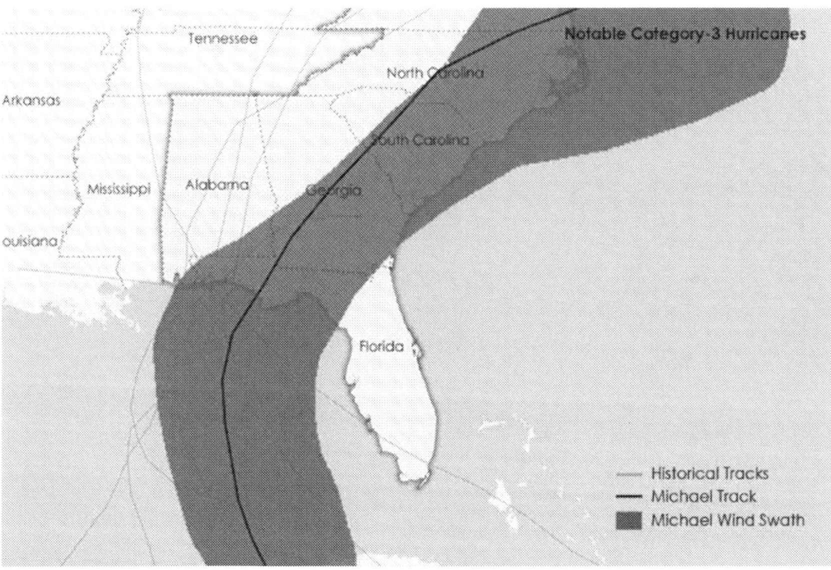

Source: NHC.

Figure 9. Notable Category 3 Atlantic Hurricanes.

Unlike Florence, Michael was a compact, fast-moving storm; this mitigated the potential of widespread damage. The Florida Panhandle did not receive nearly the same rainfall totals as experienced during Florence (2018) or Harvey (2017).

Table 2. Hurricane Michael residential and commercial loss

Peril	Residential and Commercial Total Losses in Billions (U.S. Dollars)
Wind	1 - 1.5
Flood	19 - 28.5
Total	20 - 30

Source: CoreLogic.

Pacific Hurricanes

Earlier in the season, Hurricane Lane came close to the Hawaiian Islands but did not make landfall. At its strongest, it was a Category 5 hurricane, but as it skimmed along the edge of the islands, the main impact came from the outer bands' tropical storm force winds and over 50 inches of rain.[113]

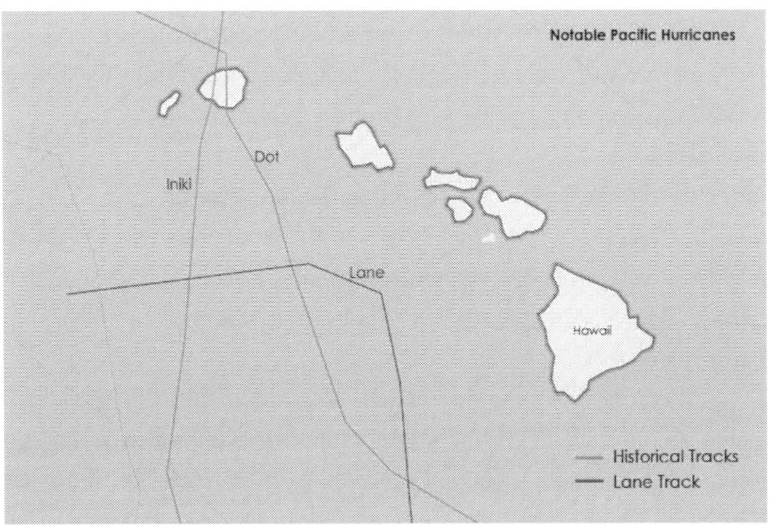

Source: NHC.

Figure 10. Notable Pacific Hurricanes.

---

[113] https://www.accuweather.com/en/weather-news/photos-lane-batters-hawaii-with-feet-of-rain-devastating-flooding-and- landslides/70005887.

As a result, the loss experienced from wind during this event was low. Annually, between four and five tropical cyclones are observed on average in the Central Pacific,[114] but landfalling hurricanes in Hawaii are historically quite rare. Hurricane Iniki (1992), was the second recorded named hurricane to make landfall in Hawaii since Hurricane Dot in 1959.

International Events

Cyclone Mekunu

In May, Cyclone[115] Mekunu made landfall in Oman as the strongest cyclone to make landfall in the Arabian Peninsula since reliable records began. This extremely severe cyclonic storm made landfall near Salalah, the capital of the Dhofar province, as a Category 3 storm with maximum sustained wind speeds of 115 mph. Significant damage from wind was reported along with widespread flooding. According to Oman's Capital Market Authority (CMA),[116] the insured losses from Mekunu reached OMR 108 million (US$281 million). Events such as this reaffirm that cyclone risk exists in the Middle East and North Africa (MENA) region; this stresses the importance of global insurers quantifying and managing their risk from potential cyclones in order to remain resilient.

Typhoon Jebi (Maymay) and Typhoon Trami (Paeng)

The Northwest Pacific Basin is one of the most active across the globe, with 26 tropical storms, 16 typhoons and nine intense typhoons occurring on average per year (1965-2017).[117] The 2018 season was above average with 30 named storms.

On September 4, Typhoon[118] a Jebi (Maymay) made landfall in Japan along the Kochi and Tokushima prefectures in Shikoku as a Category 2 typhoon and later made a second landfall west of Osaka on the island of Honshu at the same strength. Jebi was the strongest typhoon to make

---

[114] http://www.prh.noaa.gov/cphc/summaries/.
[115] http://www.aoml.noaa.gov/hrd/tcfaq/A1.html.
[116] https://cma.gov.om/Home/News/7344?page=1&year=2018.
[117] http://www.tropicalstormrisk.com/docs/TSRNWPForecastMay2018.pdf.
[118] http://www.aoml.noaa.gov/hrd/tcfaq/A1.html.

landfall in Shikoku in 25 years. The storm caused widespread damage from wind and storm surge. Kansai International Airport in Osaka Bay was inundated by flooding caused from the record high storm surge of 3.29 meters (10.8 feet).

On September 30, Typhoon Trami (Paeng) made landfall near Tanabe in Japan's southern Wakayama prefecture as a Category 2 typhoon with maximum sustained wind speeds of 109 mph. Prior to making landfall on the island of Honshu, Trami caused significant damage on the Okinawa, Kyushu and Shikoku Islands. Trami brought significant damage from wind and both storm surge and precipitation-based flooding.

Table 3. CoreLogic estimates of 2018 Japanese typhoon damage

Storm Name	Total Loss Estimate in U.S. Dollars	Total Loss Estimate in Yen
Typhoon Jebi (Maymay)	$2 billion - $4 billion	¥224 trillion - ¥450 trillion
Typhoon Trami (Paeng)	$1 billion - $3 billion	¥112 trillion - ¥338 trillion

Source: CoreLogic.

These loss estimates were determined using the CoreLogic Asia Typhoon Model[119] and evaluating proxy events (Jebi-like events) and their respective losses using the CoreLogic Insured Exposure Database (IED).

Typhoon Mangkhut (Ompong)

Super Typhoon Mangkhut (Ompong) made landfall in Northern Luzon in the Philippines as a Category 5 storm on September 14 with maximum sustained wind speeds of 167 mph. The super typhoon[120] headed towards Southwest China and made its final landfall in the Guangdong Province, west of Hong Kong, as a Category 1 storm with wind speeds of 92 mph on September 16. The insurable loss estimates for this event in China and Hong Kong were about $US250 million. The loss estimate for this event

---

[119] https://www.corelogic.com/products/asia-typhoon.aspx.

[120] http://www.aoml.noaa.gov/hrd/tcfaq/A3.html.

includes all property assets, building and contents from the residential, commercial, industrial and agricultural sectors. Government property, infrastructure, crops and livestock are not included in this portfolio. The loss estimates from this portfolio do not include loss of income and inland flooding.

Figure 11. The aftermath of Typhoon Mangkhut's devastation.

The storm track was about 100 kilometers from Hong Kong which helped mitigate the damage potential given the high property exposure in the region. Shenzen, a large Hong Kong city immediately inland does not appear to have suffered significant damage from this event. Though Mangkhut was a Category 5 typhoon at landfall in the Philippines, the insured loss was relatively low due to sparse population at the site of landfall and low insurance exposure.

Flood

1,000-Year Floods Again

Heavy rainfall throughout the U.S. contributed to billions of dollars in losses again in 2018. Much like 2016, last year's count of 1,000-year flood events was high. This level of flooding in recent years has become almost commonplace, occurring in the same spots repeatedly.

- Ellicott City, Maryland, which is still recovering from the July 2016 1,000-year flood event, was impacted in May 2018 with yet another a 1,000-year event.
- North and South Carolina also experienced severe rainfall-induced flooding from Hurricane Florence, just two years after Hurricane Matthew and three years after Hurricane Joaquin's impacts.
- In July, Houston, Texas, streets and properties flooded echoing Hurricane Harvey less than a year prior. Central Texas also experienced historic rainfall totals in September which resulted in Governor Greg Abbott issuing a disaster declaration for 18 counties as the rainfall continued into October.
- Both Northern Wisconsin and Northern Michigan experienced a 1,000-year flood. Northern Wisconsin also received a significant amount of flooding back in 2016.

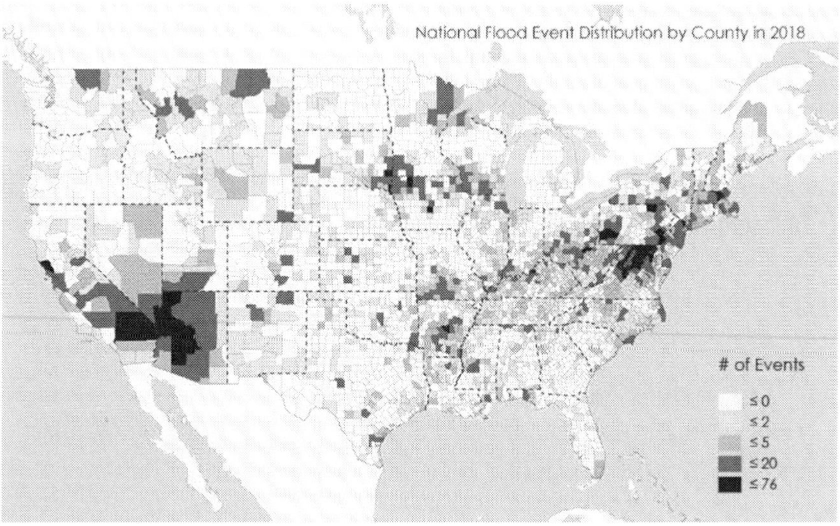

Source: CoreLogic.

Figure 12. National Flood Event Distribution by County in 2018.

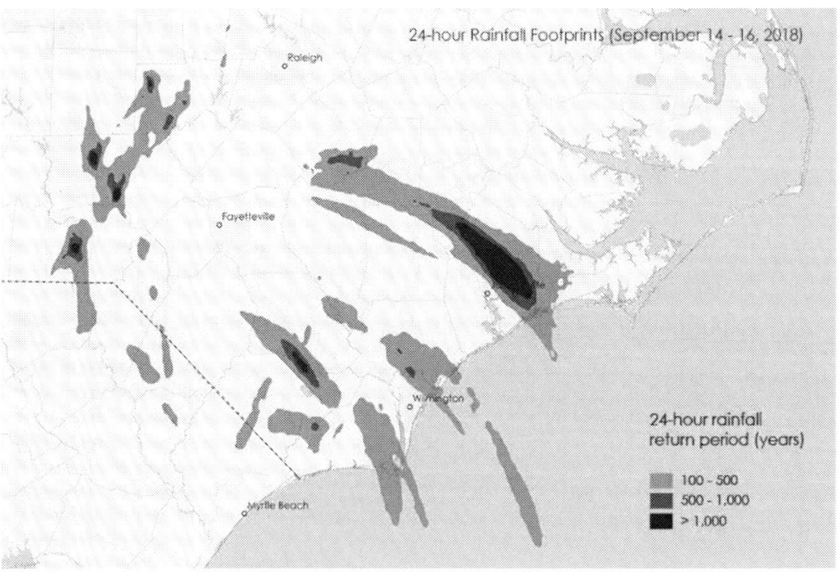

Source: CoreLogic.

Figure 13. 24-Hour Rainfall Footprints.

By comparing the losses of both Hurricanes Harvey and Florence, a similar pattern begins to take shape. Figure 5 presents 24-hour rainfall footprints for Florence between September 14 and September 16 captured by CoreLogic Weather Verification.[121] 24-hour rainfall intensities post-landfall exceeded the 1,000-year level.

Overall, in 2018, there were over 1,600 significant flood events that occurred in the United States, 59 percent of which were flash flood related. These events clearly demonstrate that 1,000-year flood events[122] should not be brushed off as an anomaly that occurs once a millennium but instead can and do repeat in as little as two years.

Hurricanes: A Contributing Cause to Severe Inland Flooding

In September 2018, Hurricane Florence trudged towards the North Carolina coast, moving at just 5 mph at its slowest point. This stalling effect mimicked Hurricane Harvey in 2017, creating the same conditions that resulted in major inland flooding from severe rainfall.

The correlation between severe inland flooding and hurricane activity has become stronger, indicating that lower category storms on the Saffir-Simpson Hurricane Wind Scale have the capacity to cause as much flood damage as stronger systems. Florence, likewise, caused flooding all along its inland path from North and South Carolina up into New Jersey.

Factors Influencing Flood Severity

It is not only the inches of rain that make a difference in damage to properties. Watersheds in the paths of rain play an important role in determining the impact as well. For instance, although Hurricane Lane brought more rainfall to the Hawaiian Islands than Hurricane Florence brought to North Carolina, Florence caused more inundation. In Hawaii, the land and river channel slopes are steeper than those in North Carolina, and this topography allows flood waters to flow to the ocean within several hours, significantly reducing storm water accumulation, inundation areas

---

[121] https://www.corelogic.com/solutions/weather-verification-services.aspx.
[122] http://www.nws.noaa.gov/oh/hdsc/aep_storm_analysis/.

and flooding duration. As a result, only 4.24 percent of the land in Hawaii is a designated SFHA compared to North Carolina's 20.11 percent.

That said, flooding is not limited to the SFHAs and can expand beyond the designated zone. Although natural rivers have the capacity to handle smaller, more frequent flood events, extreme inundation can reach much broader areas outside of the 100-year floodplain.

The national average of landmass in an SFHA is roughly 12 percent (Figure 7). Two coastal states with large relatively flat land areas, Louisiana and Florida, are the national leaders in this ratio, ranging from 45 percent to 55 percent. Comparatively, the ratio for Texas, which has seen several extreme events in the past few years, is about 15 percent. SFHAs do not cover the full spectrum of flood risk – three of four flood claims made in Houston occurred outside of the SFHA.

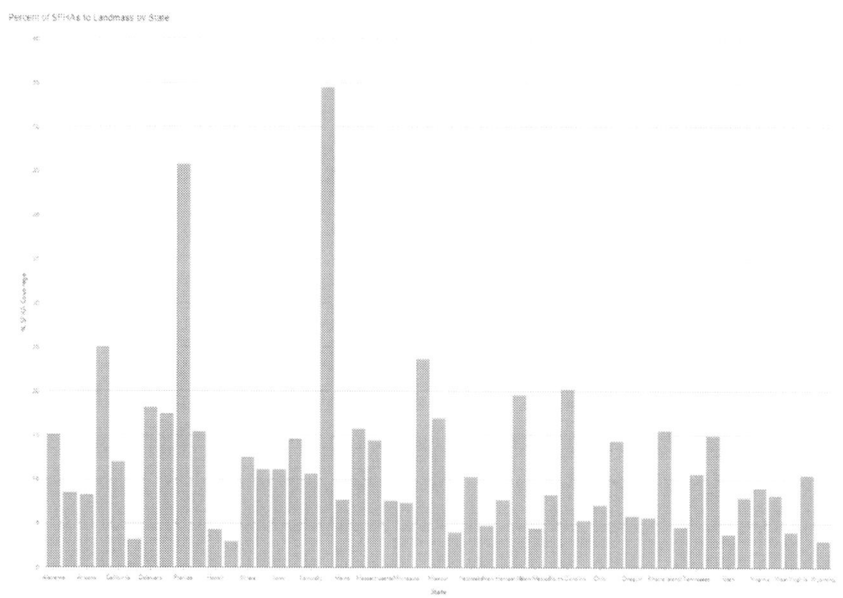

Source: CoreLogic Analysis on the Percent of SFHAs to Landmass by State.

Figure 14. Percent of SFHAs to Landmass by State.

Based on CoreLogic national property analysis, 6% of properties nationwide are within SFHAs. According to a recent Wharton report,[123] unfortunately, only about one-third of those have National Flood Insurance Program (NFIP) flood insurance policies - a stunning indication that the majority of properties in the U.S. don't have sufficient financial protection from a flood event.

Based on our Florence flood loss analysis utilizing the CoreLogic U.S. Inland Flood Model,[124] 59 percent of affected properties for the event were outside of a Federal Emergency Management Agency (FEMA) designated SFHA. The estimation of Florence's flood losses by state is presented in the table below, and you can read the full analysis here.[125]

Table 4. Residential and commercial insured and uninsured (storm surge and inland) flood loss estimation for Hurricane Florence ($ billion)

State	Insured Flood Loss	Uninsured Flood Loss
North Carolina	4.5 – 7.5	10 – 14.5
South Carolina	1 - 2	2.5 – 3.5
Virginia	~ 0.5	~ 0.5
Total	6 - 10	13 – 18.5

Source: CoreLogic.

As the most common natural disaster in the U.S., flooding is an event for which insurers, homeowners and local governments should adequately prepare. Extreme rainfall and slow-moving storms continue to create losses year after year. 1000-year flood events are seemingly the new normal, and we expect billions of dollars in losses in the years to come.

123 https://riskcenter.wharton.upenn.edu/wp-content/uploads/2018/07/Emerging-Flood-Insurance-Market-Report.pdf.

124 https://www.corelogic.com/products/us-inland-flood-model.aspx.

125 https://www.corelogic.com/news/the-aftermath-of-hurricane-florence-is-estimated-to-have-caused-between-20-billion-and-30- billion-in-flood-and-wind-losses-cor.aspx.

Wind

According to CoreLogic wind analysis, 1,702,726 square miles or 56 percent of the continental U.S. was impacted by severe (>60 mph) wind gusts in 2018. On the state level, increased thunderstorm activity and wind related to Hurricane Florence were responsible for an above-average year in both Colorado and North Carolina.

Source: CoreLogic.

Figure 15. Total Land Area Affected by Wind (>60 mph) by State in Acres.

When analyzing very severe wind gusts (>80 mph), 21,655 square miles, or 0.71 percent of the continental U.S., were impacted in 2018, slightly more than the previous two years. Although not nearly as big as Hurricanes Matthew and Irma, Hurricane Michael's severe winds reached nearly a quarter of the total area impacted by winds in 2018. Michael's 100 mph wind gusts extended up to 85 miles inland.

Overall, the strongest wind gusts associated with severe thunderstorms in 2018 occurred on June 28 in central North Dakota. CoreLogic estimated wind gusts of up to 120 mph over mainly rural areas. The strongest wind gusts of the year were associated with Hurricane Michael with estimated winds of at least 155 mph near Mexico Beach off the coast of Florida.

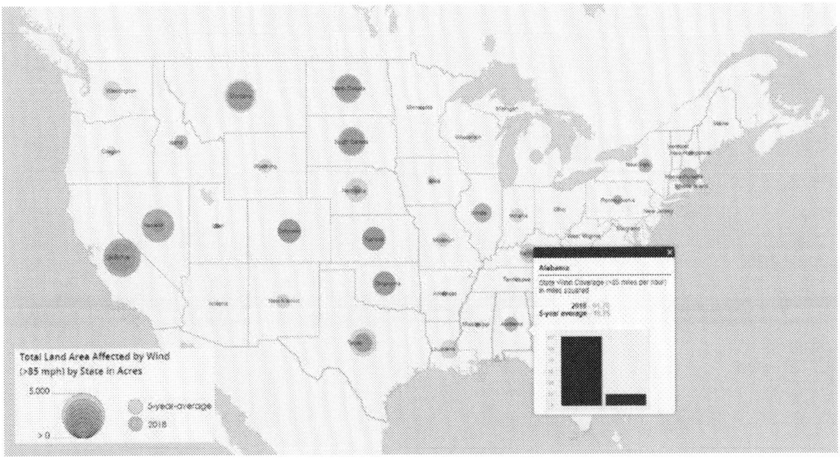

Figure 16. Total Land Area Affected by Wind (>85 mph) by State in Acres.

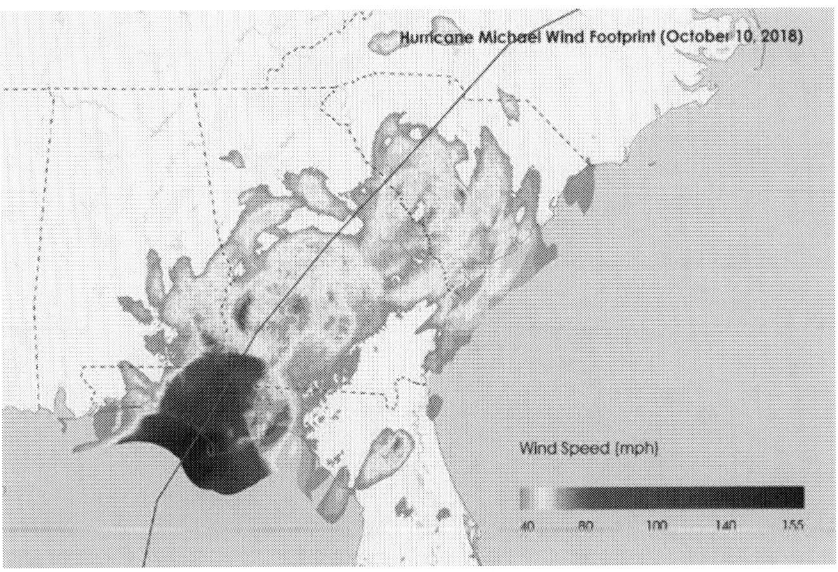

Source: CoreLogic.

Figure 17. Wind Footprint from Hurricane Michael.

Thunderstorms

On June 28, North Dakota experienced over 500 severe wind reports from three separate lines of thunderstorms, or squall lines. The first and most severe line of storms stretched from western North Dakota to northern Minnesota while the other two stretched along the Missouri and Illinois border, down through Mississippi and Alabama. Notable cities impacted include Bismarck, North Dakota; Fargo, North Dakota; St. Louis, Missouri; and Birmingham, Alabama.

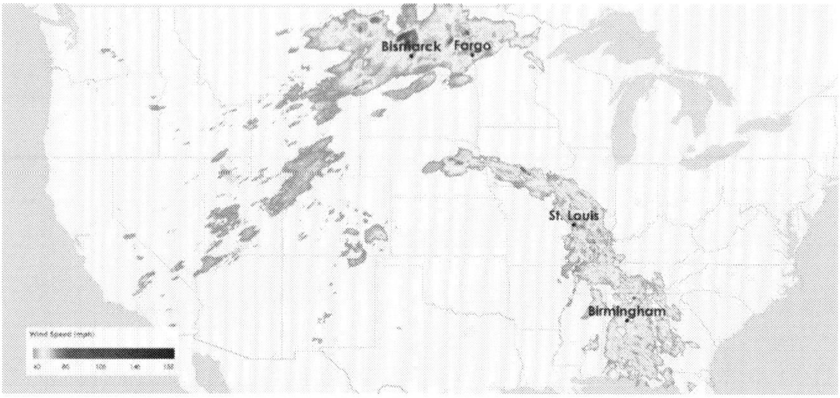

Source: CoreLogic.

Figure 18. June 2018 Wind Footprint.

Source: CoreLogic.

Figure 19. May 2018 Wind Footprint.

With almost 500 severe wind reports, the second largest event occurred on May 14 and 15 with two separate, severe squall-lines impacting many of the largest cities on the East Coast, including Washington D.C.; Baltimore, Maryland; Philadelphia, Pennsylvania; New York, New York; and Boston, Massachusetts.

As we move into early 2019, weak El Niño conditions, or the warming of waters in the equatorial Pacific, are expected to persist. Recent research suggests that when El Niño conditions are present, average to below average thunderstorm activity from the Plains into the Southeast U.S. can occur in late winter and into the spring.

Wildfire

While the total number of acres and quantity of fires in the U.S. did not quite match 2015-2017 levels, 2018 was nonetheless a devastating year for wildfires.

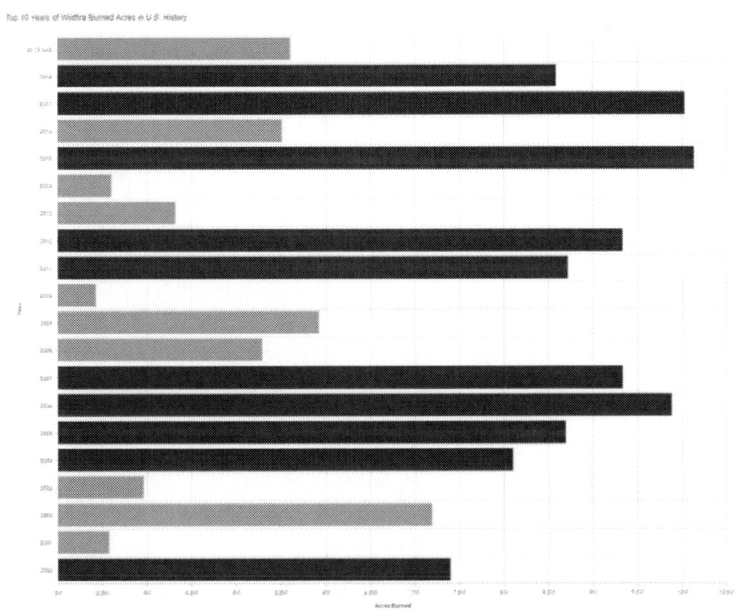

*Total Acreage Reported As Of December 28, 2018.
Source: National Interagency Fire Center, 2018.

Figure 20. Top 10 Years of Wildfire Burned Acres in U.S. History.

The number of acres that burned in 2018 was the eighth highest in U.S. history as reported through December 28, 2018. All of the top 10 fires have occurred since the year 2000, five of which have occurred since 2010, and all of which have exceeded the 30-year average. A total of 11 western states had at least one wildfire that exceeded 50,000 burned acres (Table 5), indicating where the wildfire risk was high given prevailing conditions and opportunity for ignition.

Table 5. Western states with at least one wildfire greater than 50,000 acres in size in 2018

State	# fires > 50k acres
California	7
Colorado	3
Idaho	4
Nevada	4
New Mexico	2
Oklahoma	2
Oregon	7
Texas	1
Utah	3
Washington	3
Wyoming	1

*As of November 2018.
Source: Geomac Wildland Fire Support, Fire Perimeters, 2018.

Due to a combination of high winds and dry conditions in populous areas, California was again the victim of late season wildfires. 2018 broke the record for the largest fire, deadliest fire and most destructive fire in the state's history. The state also suffered more property loss from wildfires than any other state in 2018.

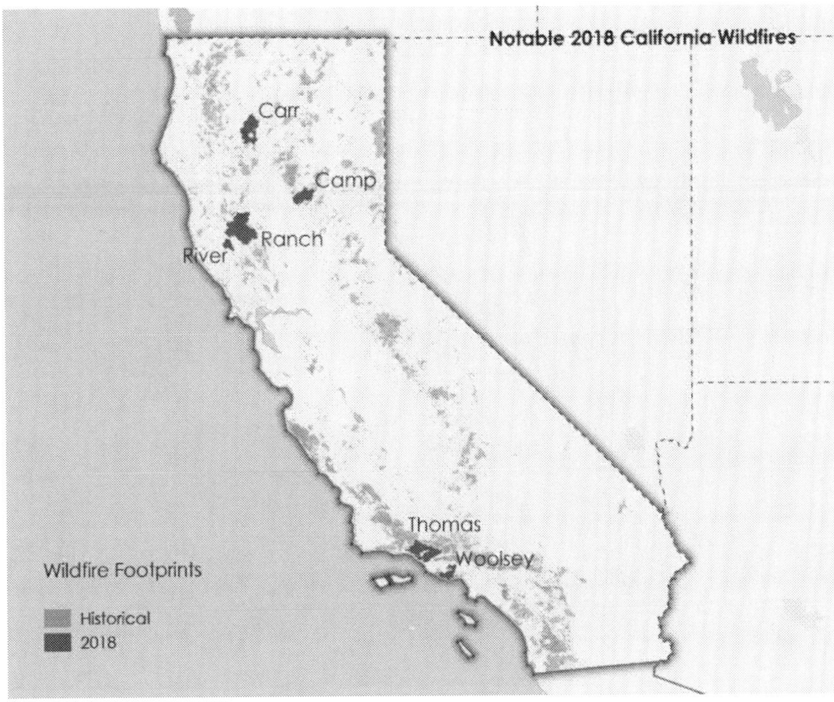

Source: CalFire, GeoMAC, USGS, Esri.

Figure 21. Wildfire Footprint across California, Including Historical and 2018 Wildfires.

Major Fires in 2018

After record-setting destruction by the Tubbs, Nuns and Atlas fires just one year prior, Northern California experienced another devastating blow when the Camp Fire[126] destroyed nearly the entire city of Paradise and damaged or destroyed over 18,804 structures in November.

---

[126] http://cdfdata.fire.ca.gov/incidents/incidents_details_newsreleases?incident_id=2277.

At the same time, the Woolsey Fire[127] in the coastal community of Malibu destroyed more than 1,600 structures[128] and caused tens of thousands of people to evacuate. CoreLogic estimates that the combined total loss for these two wildfires is between $15 billion and $19 billion.[129]

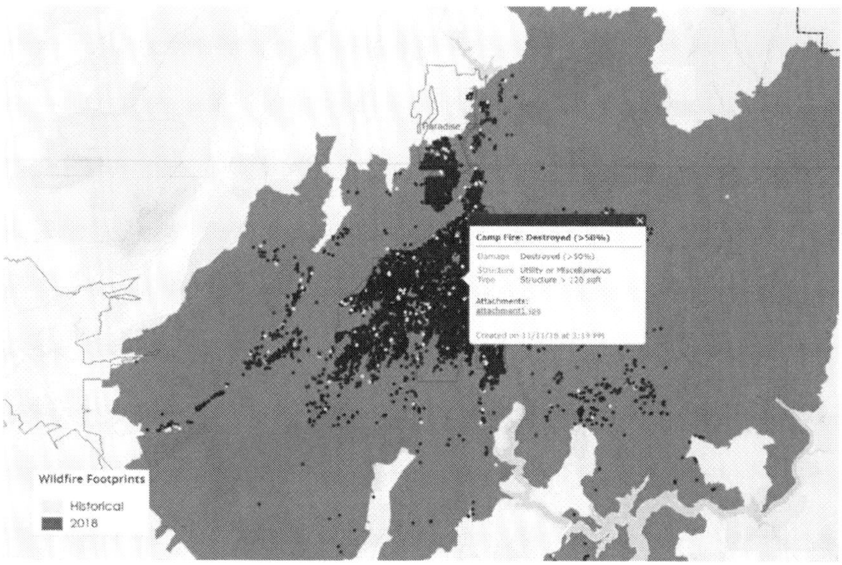

Figure 22. Camp Fire Footprint.

The amount of properties destroyed by the Camp Fire is more than three times the number of structures destroyed in 2017's Tubbs Fire. The loss of life attributed to the Camp Fire (85 lives) is also more than triple the three next highest fires combined. The large loss of life is partially attributed to the speed at which this fire consumed the city of Paradise.

The Woolsey Fire, with 1,643 destroyed structures is currently ranked as the seventh most destructive wildfire in California history, just ahead of the 2018 Carr Fire.

---

[127] http://cdfdata.fire.ca.gov/incidents/incidents_details_info?incident_id=2175

[128] https://www.fire.ca.gov/current_incidents/incidentdetails/Index/2282.

[129] https://www.corelogic.com/news/the-camp-and-woolsey-wildfires-in-california-cause-devastating-losses-between-15-billion-and-19-billion-according-to-corelogic.aspx.

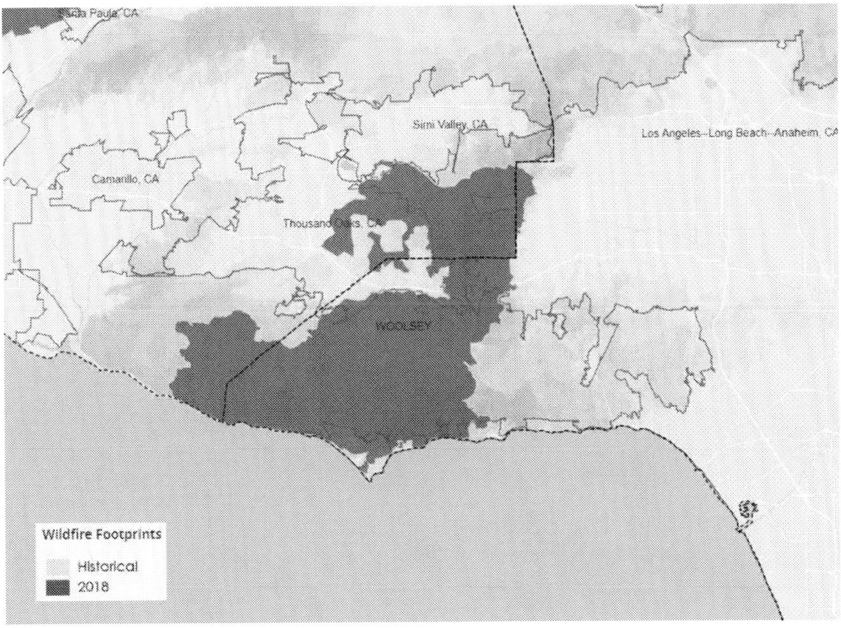

Figure 23. Woolsey Fire Footprint.

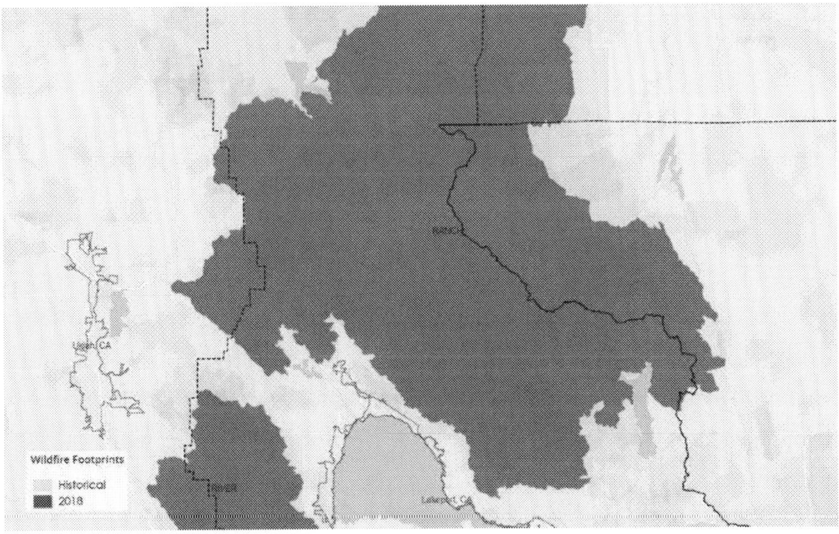

Figure 24. Mendocino Complex Fire Footprint.

Earlier in the season, the Mendocino Complex Fire[130] was a combination of the Ranch and River Fires covering parts of Glenn, Colusa, Lake, and Mendocino Counties. The fire burned from July to September and grew to 459,123 acres, making it the largest wildfire in California history – displacing the record set by 2017's Thomas Fire by more than 175,000 acres. In the process, it destroyed 280 structures.

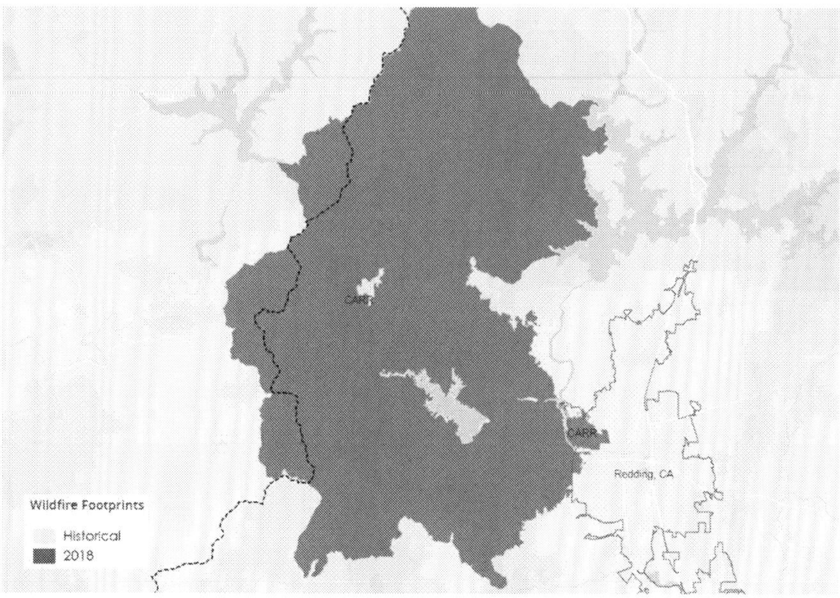

Figure 25. Carr Fire Footprint.

The Carr Fire[131] in Shasta and Trinity Counties was approximately half the size of the Mendocino Complex Fire but was far more destructive, destroying 1,604 structures.

The Carr Fire burned into the city of Redding, causing thousands to evacuate nearby areas. It also caused a fire tornado or fire whirl, a rather uncommon occurrence in most wildfire events. Measured at 143 miles per hour, the equivalent of an EF3 tornado, the rotating fire and wind field

---

[130] http://cdfdata.fire.ca.gov/incidents/incidents_details_info?incident_id=2175.
[131] http://cdfdata.fire.ca.gov/incidents/incidents_details_info?incident_id=2164.

added to the damage of structures and infrastructure within the wildfire perimeter.

2018 Compared to the Previous 10 Years

Recent wildfire activity in the U.S. has continued to be extensive and costly, with three of the last four years resulting in more than 8 million acres burned[132] and tens of thousands of structures destroyed. While the number of fires and the total acreage are an important aspect of fire risk, the location of these events in relation to urban areas and other clustered developments is the most crucial factor in determining the damage that can result from this hazard.

Wildfires are distinctive because unlike many other perils, they frequently cause total loss of the structure and its contents. It's critical to have an accurate understanding of the reconstruction cost[133] associated with at-risk homes and businesses; different from the market value of a home, reconstruction cost is an analysis of the materials and labor it takes to completely rebuild a home from the ground up. It's recommended to reevaluate this every two years, even if using a factor or index, to not only understand what is at stake but to manage the capital necessary to restore a homeowners' livelihood after a catastrophe.

Earthquake

While there wasn't a major damaging earthquake in the U.S. in 2018, the level of earthquake activity was not abnormal. According to the U.S. Geological Survey (USGS), as of December 20, a total of 470 earthquakes with a magnitude of 3.0 or greater occurred in the conterminous U.S. in 2018.

---

[132] https://www.nifc.gov/fireInfo/nfn.htm.

[133] https://www.corelogic.com/landing-pages/rct-the-benchmark-of-reconstruction-valuation.aspx.

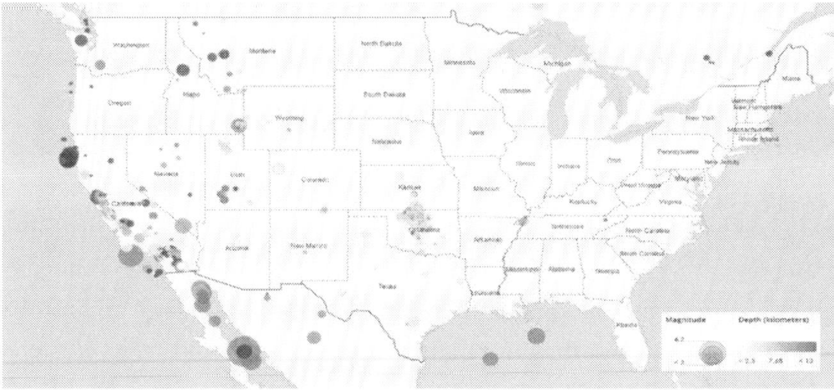

Source: USGS.

Figure 26. 2018 Earthquakes (Magnitude 3.0 or Greater).

The earthquake activity in Oklahoma was a noteworthy change in 2018. For nearly the past decade, the number of earthquakes in Oklahoma has garnered the attention of scientists and risk managers across the country. Unlike California, which has a long history of earthquake activity, historically, Oklahoma had experienced very low earthquake activity until 2009 when the rate began to rapidly increase. Research suggests[134] this increase is likely correlated with increased oil and gas activity, specifically the pumping of waste water at fluid injection wells. 2015 saw the peak of earthquake activity in Oklahoma with the number continuing to decline in the past three years. The rate of earthquake activity increased in 2018 compared to 2017, but for the first time since 2014 there were a greater number of earthquakes in California than Oklahoma once again (Figure 20).

In 2018, less than 20 earthquakes of magnitude 3.0 and greater were reported in Idaho, approximately 10 times less than the swarm of events near Soda Springs in 2017.

[134] https://earthquake.usgs.gov/research/induced/myths.php.

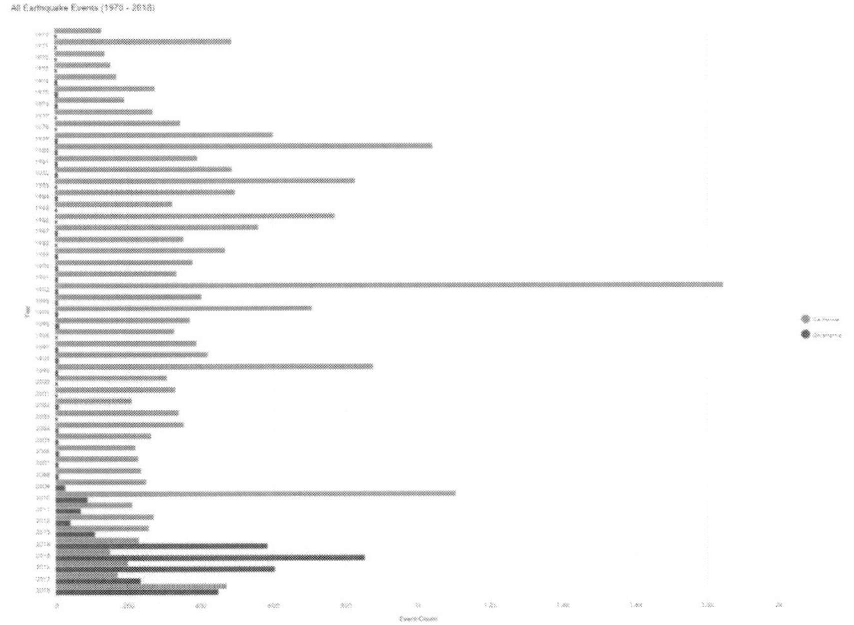

Source: USGS.

Figure 27. Oklahoma vs. California: Number of Earthquakes Per Year Since 1970.

Alaska Earthquake (M7.0)

Outside of the continental U.S., the largest and most damaging earthquake of the year was a magnitude 7.0 quake near Anchorage, Alaska,[135] on November 30. The event was widely felt across Alaska, and ground motions impacted a very large area, but the damage was concentrated in close proximity to Anchorage. In the days following the initial earthquake, over 1,000 aftershocks were recorded. CoreLogic analysis indicated that 28,733 homes were within the USGS ShakeMap designated area of Modified Mercalli Intensity (MMI) VII, or susceptibility to very strong ground shaking, which is typically where structures begin to see damage.

---

[135] https://earthquake.usgs.gov/earthquakes/eventpage/us1000hyfh/executive.

Hokkaido Earthquake (M6.6)

On September 5, a magnitude 6.6 earthquake[136] struck 27 km east of Tomakomai at a depth of 35 km on the island of Hokkaido in Japan. This occurred as a result of shallow reverse faulting in which the earth is forced together by compression. The epicenter was approximately 60 km south-southeast of Sapporo, the prefecture capital, while shaking was felt throughout the region. More than 41 fatalities[137] were recorded, and according to CoreLogic, physical losses to the agricultural industry were in the region of US$400 million. Notably, much of the damage occurred due to landslides, and destroyed infrastructure led to longer than expected downtime in business operations, resulting in increased business interruption losses.

Sulawesi/Palu Earthquake and Tsunami (M7.5)

On September 28, a magnitude 7.5 earthquake[138] struck 70 km north of Palu at a depth of approximately 10 km, triggering a tsunami on the Indonesian island. A total of seven earthquakes were recorded within a seven-hour time frame. This included two magnitude 5.7 earthquakes and one magnitude 5.8 earthquake closer to the populous area of Palu less than 20 minutes after the main shock event, around the same time the tsunami hit the coastal areas of Palu. The devastation caused 2,081 fatalities[139] and destroyed more than 68,451 houses.[140] The destruction of homes and loss of lives was due to both the tsunami on the coast and the liquefaction of softer soils further inland which consumed whole villages[141] in the areas of Petobo and as far south as Sidera.

---

[136] https://earthquake.usgs.gov/earthquakes/eventpage/us2000h8ty/executive#executive.

[137] http://www.asahi.com/ajw/articles/AJ201809110026.html.

[138] https://earthquake.usgs.gov/earthquakes/eventpage/us1000h3p4/executive.

[139] https://ahacentre.org/situation-update/situation-update-no-15-sulawesi-earthquake-26-october-2018/.

[140] https://ahacentre.org/situation-update/situation-update-no-15-sulawesi-earthquake-26-october-2018/.

[141] https://earthobservatory.nasa.gov/images/92836/devastation-in-palu-after-earthquake-tsunami.

Volcano

Mt. Kilauea Volcanic Eruption

Hawaii dominated the news in 2018 with the eruption of Mt. Kilauea. The island of Hawaii, also known as 'the Big Island', is made up of five shield-type volcanoes where explosive eruptions are uncommon. Typical eruptions for shield volcanoes contain lava that is basaltic in nature, with a low viscosity. This type of lava is runnier and travels faster than in other types of volcanic eruptions with higher viscosity lava.

On May 17, Kilauea, the most active of the five volcanoes, experienced an explosive eruption. The preceding and subsequent basaltic lava flow rose to the surface via a recorded 24 fissures (or vents). Shield volcanoes experience these uncommon explosive eruptions when they interact with water deposits that get into the vent. This episode of explosive eruption was preceded and proceeded by earthquake activity. An identified magnitude 6.9 main shock earthquake in the Puna region occurred on May 4 with the largest foreshock of magnitude 5.4 approximately an hour earlier and numerous aftershocks continuing for several weeks.

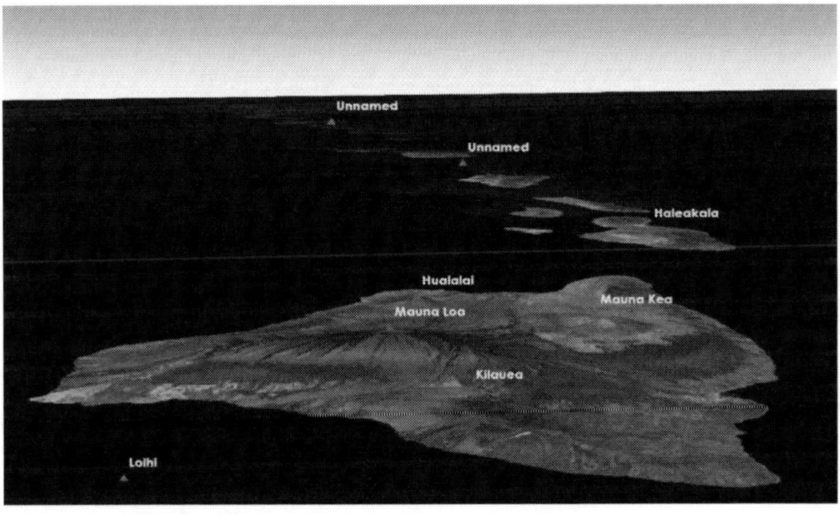

Figure 28. Hawaii Volcanoes.

Figure 29. The 2018 Kilauea eruption was just one episode of a larger event, as the volcano had been continually erupting since 1983.[142]

Throughout the duration of the eruption, 1,700 people were evacuated from homes within the thermal zone defined by the USGS. The Puna district, a less populous region compared to the cities of Kona and Hilo, was impacted by the eruption the most, with the heaviest damage reported in Leilani Estates and Lanipuna Gardens.

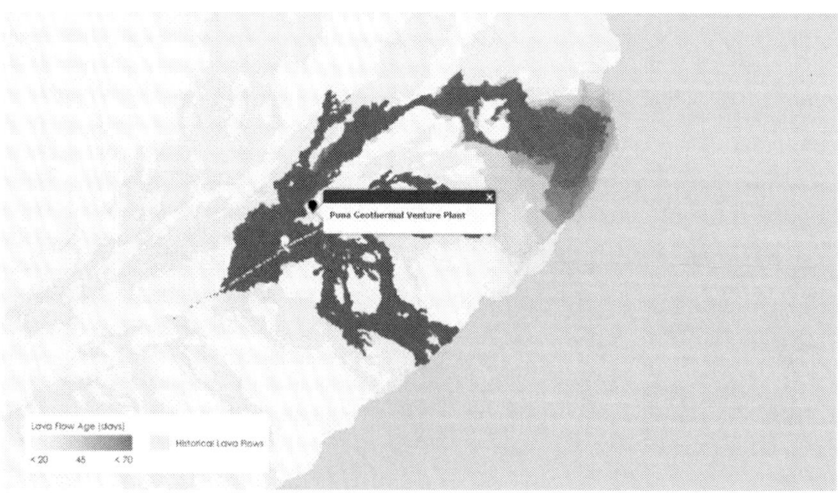

Figure 30.

[142] https://volcanoes.usgs.gov/volcanoes/kilauea/geo_hist_1983.html.

The lava covered 13.7 square miles (35.5 square kilometers), and 657 homes are confirmed to have been destroyed. A notable impact of this event is the closure of the Puna Geothermal Venture (PGV) Plant, a geothermal power plant providing roughly a quarter of Hawaii's energy supply. It was capable of producing 38 million watts worth an estimated $13 million in revenue per annum,[143] with an estimated building value of around USD $170 million.[144] Damage had been reported to the substation and transmission lines. Two geothermal wells had been covered in lava where the extent of the damage is unclear. Since May 22, the plant has remained shut down and non-operational through December 2018. Fortunately, backup power supply for the island was provided by diesel power generators. As a result, the island did not experience a power shortage during the event.

By mid-August the thermal zone had reached its fullest extent and had stopped expanding.

Hail

Overall, hail activity in 2018 was below the CoreLogic calculated average (2009-2017). According to CoreLogic, 134,167 square miles, or 4.4 percent of the continental U.S., were impacted by severe hail which the National Weather Service (NWS) defines as 1.0 inch or greater.

Some of the largest hail events of the year occurred relatively early in the season. On March 19 in Cullman, Alabama, and July 20 in Thompkinsville, Kentucky, CoreLogic estimated that hail over 4 inches in diameter had fallen.

---

[143] https://www.hawaiianelectric.com/billing-and-payment/rates-and-regulations/average-price-of-electricity.

[144] https://investor.ormat.com/file/Index?KeyFile=393711325.

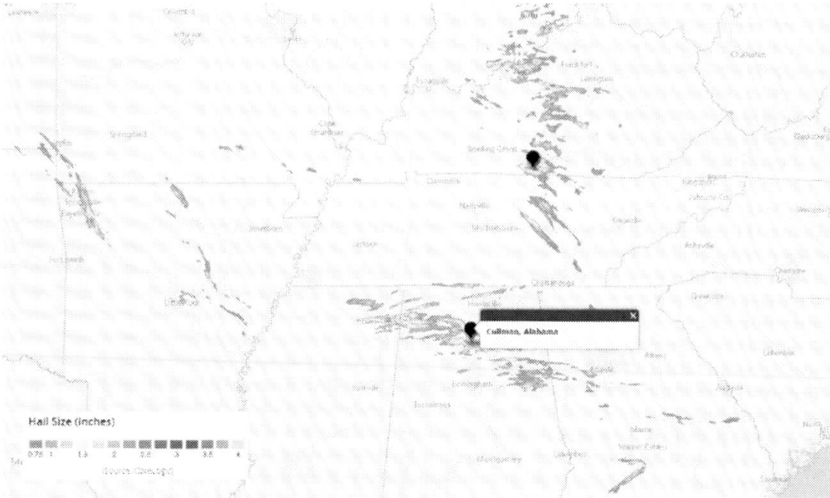

Source: CoreLogic.

Figure 31. Hail Event in Cullman, Alabama.

2018 Compared with Previous Years

The total area impacted by severe hail in 2018 was the lowest observed since at least 2009, down 20 percent from the 2009-2017 average.

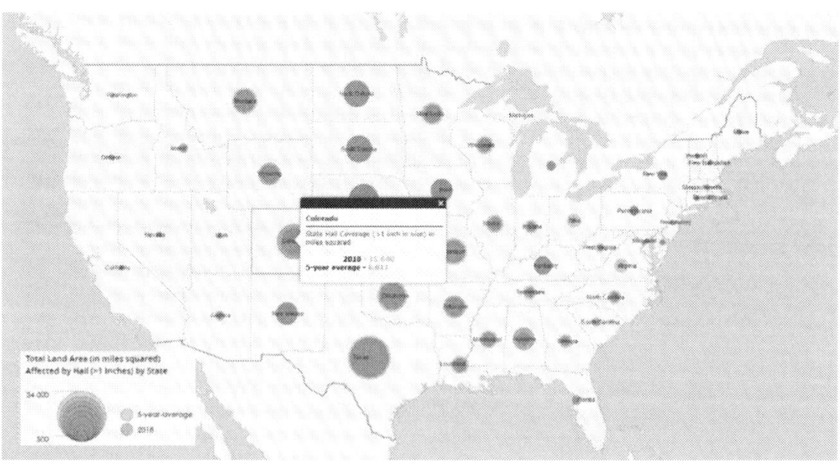

Source: CoreLogic.

Figure 32. Total Land Area (In Miles Squared) Affected by Hail (>1.0 Inch) by State.

However, the High Plains in Colorado and Wyoming and portions of the Southeast received over double their five-year average total area of 1-inch hail.

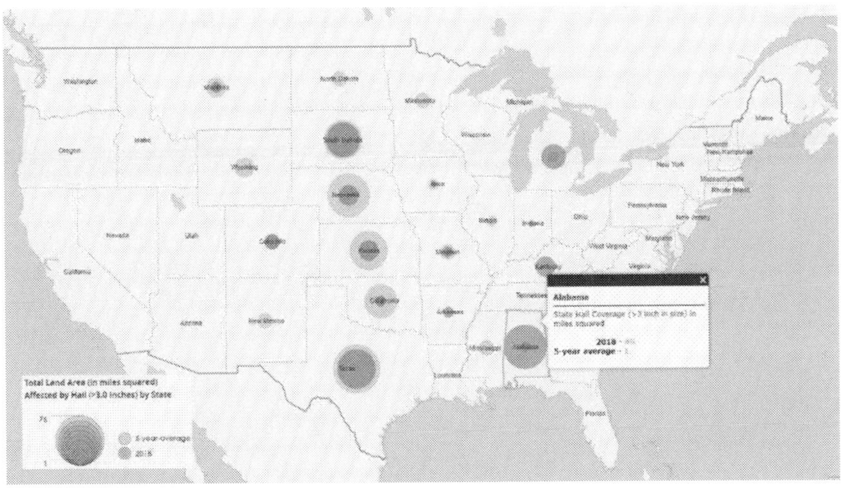

Source: CoreLogic.

Figure 33. Total Land Area (In Miles Squared) Affected by Hail (>3.0 Inches) by State.

When analyzing the area of very large hail (> 3.0 inches), activity in 2018 was an astounding 44 percent below the CoreLogic calculated average (2009-2017), with only 2016 having a lower total area. In a rare occurrence, Alabama received greater than 3.0 inch hail, the most out of any state in the country, with most of it falling the same day Cullman, Alabama, was hit.

State/Metro Areas with the Most Activity and Damage Costs

In the early morning hours of June 6, a severe hailstorm with baseball-sized hail struck the middle of the Dallas-Fort Worth metro area. The cities of Carrollton and Arlington were hit with the biggest hail, causing damage upwards of $1 billion according to the firm Karen Clark & Co, making it the third costliest hailstorm in Texas history. The Insurance Council of Texas estimated large hail damaged upwards of 20,000 structures and

25,000 vehicles. Since the storm occurred overnight, few cars were on the road; if it had occurred during the afternoon or evening, the number of damaged vehicles would have been substantially higher.[145]

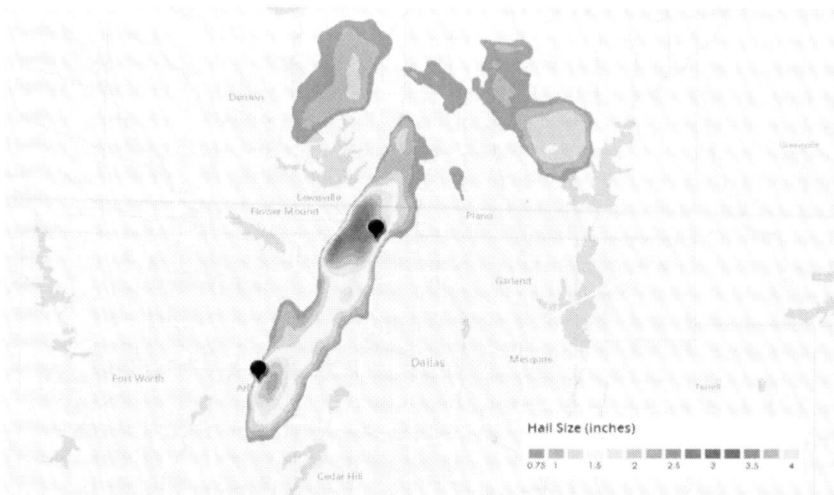

Source: CoreLogic.

Figure 34. Hail in Inches in the Dallas-Fort Worth Metro Area.

After experiencing a record hail season in 2017, Colorado saw severe hailstorms strike Denver and the Colorado Springs area twice in 2018. On June 18, a series of hailstorms hit northern Denver suburbs along the Front Range, including the city of Boulder. The Rocky Mountain Insurance Information Association (RMIIA) estimated upwards of $276 million in damage, $191 million of which came from auto claims. Meanwhile on June 13 and August 6, up to baseball-sized hail hit the area south of Colorado Springs, causing $169 million and $172 million in damage.[146]

---

145 https://www.insurancejournal.com/news/southcentral/2018/07/06/494201.htm.
146 https://www.insurancejournal.com/news/west/2018/06/28/493577.htm.

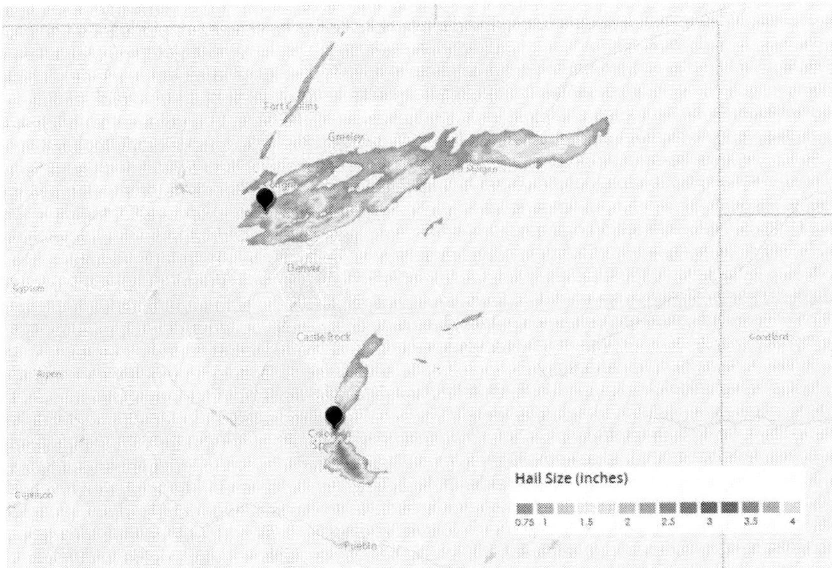

Source: CoreLogic.

Figure 35. Hail in Inches in Denver and the Colorado Springs Area.

Research suggests that hail is also affected by weak El Niño conditions. As the waters warm in the equatorial Pacific, average thunderstorm activity from the Plains into the Southeast U.S. usually occurs from late winter into spring. If El Niño conditions persist, average storm and hail activity can be expected across these areas well into spring 2019.

Tornado

Overall tornado activity in 2018 was 15 percent below the 10-year average.[147] According to the Storm Prediction Center (SPC), a federal entity under NOAA, there were 1,154 tornadoes through December. Iowa led the nation in the number of annual tornadoes with 84, while Texas, which usually leads with an average of 155 tornadoes a year, only had 50 tornadoes in 2018. While tornado season got off to a slow start at about 42 percent below average through May, the overall tornado activity was well

---

[147] https://www.spc.noaa.gov/climo/online/monthly/2018_annual_summary.html.

above average during the late summer and fall; still, the 2018 total is well below the totals recorded in previous years.

State/Metro Areas with the Most Activity and Damage Costs

Three major tornado outbreaks hit the Southeastern United States in 2018, which is common for a La Niña year as cold air from Canada surges southward and strengthens weather systems in the spring. The first occurred on February 24 in the Ohio Valley where 27 confirmed tornadoes were reported and three people were killed. The strongest tornado, an EF2, occurred in Northeast Arkansas and was on the ground for 42 miles.[148]

The second outbreak encompassed 27 confirmed tornadoes on March 19 across Mississippi, Alabama and Georgia, including an Atlanta suburb. The strongest tornado, an EF3 with winds up to 150 mph, hit Jacksonville State University and caused extensive damage; at its maximum, the tornado had a path width of over a mile.[149]

The third and largest tornado outbreak of the year occurred from the morning of April 13 until the evening of April 15, stretching from Western Louisiana and Arkansas all the way down to Southern Florida and up to Western Virginia. In total, 82 tornadoes were confirmed and impacted the cities of Shreveport, Louisiana and Greensboro, North Carolina. The strongest tornado, an EF3, hit Lynchburg, Virginia.[150]

In September, Hurricane Florence spawned at least 48 tornadoes across the mid-Atlantic as the system stalled out in North Carolina. One of the tornadoes near Richmond, Virginia resulted in a fatality.[151]

Like hail and wind, tornadoes are among the perils affected by weak El Niño conditions. However, the warm equatorial Pacific waters can bring about decreased tornado activity in the Plains and southeast U.S. If this continues, decreased tornado activity can be expected in those areas in the spring of 2019.

---

[148] https://www.weather.gov/pah/2018_Feb24_EventSummary.

[149] https://www.weather.gov/bmx/event_03192018.

[150] https://www.weather.gov/shv/event_2018-4-13_tornadoes.

[151] https://www.richmond.com/weather/remnants-of-hurricane-florence-gave-the-richmond-area-its-first/article_a4828a8c-cfc1-539f-a0de-4d797dd1ddd8.html.

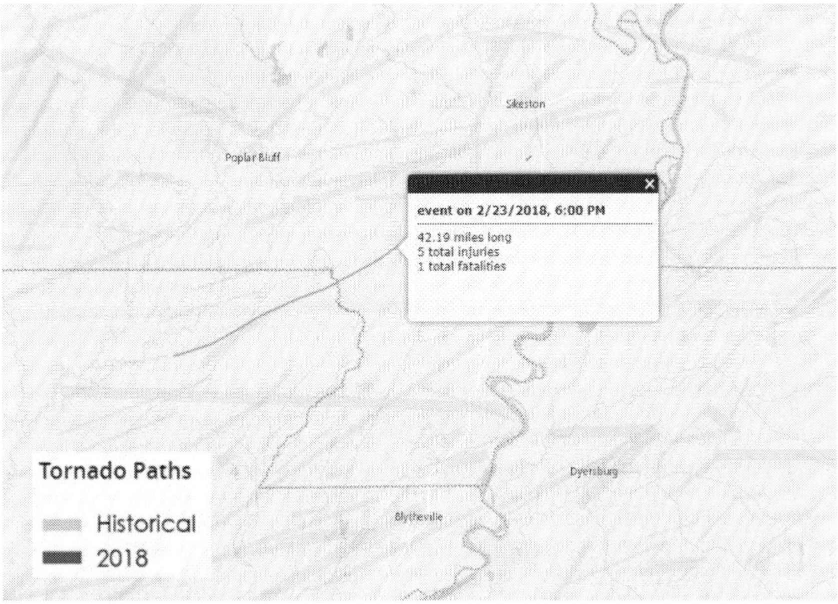

Source: National Weather Service Damage Toolkit.

Figure 36. Tornado Path for EF2 Tornado in Northeast Arkansas.

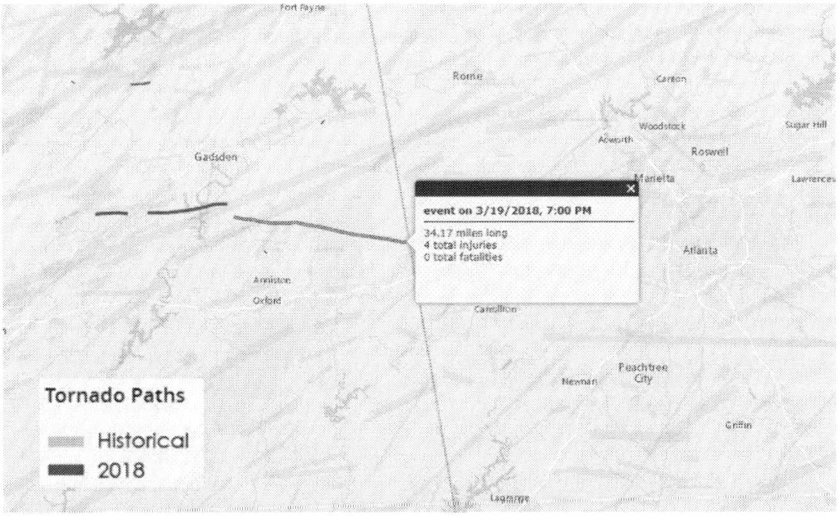

Source: National Weather Service Damage Assessment Toolkit.

Figure 37. Tornado Path for EF3 Tornado in Atlanta Area.

About CoreLogic

CoreLogic (NYSE: CLGX) is a leading global property information, analytics and data-enabled services provider. The company's combined data from public, contributory and proprietary sources includes over 4.5 billion records spanning more than 50 years, providing detailed coverage of property, mortgages and other encumbrances, consumer credit, tenancy, location, hazard risk and related performance information. The markets CoreLogic serves include real estate and mortgage finance, insurance, capital markets, and the public sector. CoreLogic delivers value to clients through unique data, analytics, workflow technology, advisory and managed services. Clients rely on CoreLogic to help identify and manage growth opportunities, improve performance and mitigate risk. Headquartered in Irvine, Calif., CoreLogic operates in North America, Western Europe and Asia Pacific. For more information, please visit corelogic.com.

White Paper Submitted by Representative Kendra Horn

How Tornado Technologies Work

Introduction

Every year tornadoes present significant risk to the population and infrastructure of the Continental United States. Understanding and assessing the damage likely caused by tornadic winds on building performance, and local infrastructure impacted, is critical for public safety and restoring those impacted. Being able to quickly distinguish who and what potentially has been impacted is crucial for providing immediate assistance from first responders, local agencies and insurance carriers that are dedicated to restoring these communities.

Like hail and wind storms, tornadoes contribute significantly to overall insurance losses annually from very small local events to very large catastrophic events. While hundreds to thousands of tornadoes are typically observed in the U.S. each year, a single tornado that hit Joplin, Mo. on May 22, 2011 caused an estimated $2.1 billion (2011USD) in insured losses according to the Insurance Information Institute.

Defining exactly where on the ground tornado damage has impacted properties and assets is critically important to understand exactly who and what may have suffered damage. Equally important is getting an accurate assessment as quickly as possible once the event is over. As part of CoreLogic's dedication to developing accurate weather verification products and services to meet the needs of the Insurance Community, it has developed Tornado Path Maps which provides the critical information needed to mobilize catastrophe managers, claims adjusters, and support personnel within minutes to hours of a tornadic event.

Similar to the use of Hail Size Maps and Wind Speed Maps from CoreLogic it is now possible to use Tornado Path Maps to improve policyholder experiences during catastrophe events. With this new technology, it will ultimately allow carriers to increase satisfaction of the insured party through faster service, and to provide emotional assurances to their policy holders that they are aware and responding to their needs in such a traumatic time.

The Challenge

Unlike hail and wind, the rapid evolution and small size of a tornado, compared to the parent thunderstorm, present unique challenges to automatically detecting the path. Most observers providing tornado reports are not located within the path and only provide visual descriptions of the tornado's appearance, which are often uncorrelated with tornadic intensity. Weather radars can observe a storm's rotation, but many strong rotation signatures are not associated with tornadoes and some tornadoes can be associated with weak rotation signatures, making it difficult to differentiate the tornadoes.

One of the most violent phenomena on earth, tornadoes are capable of completely destroying one house while leaving another untouched only a few blocks away.

With the proprietary Tornado Verification Science from CoreLogic, carriers can better understand the impact of each unique storm by comparing detailed storm maps and reports with their book of business. This enables a more proactive approach to claims management. By being able to accurately verify and pinpoint affected areas, insurance carriers can develop swifter, more targeted response plans to improve customer satisfaction, catastrophe response efficiency, and confidently detect fraudulent or unrelated claims.

Purpose

The goal of this paper is to explain how the proprietary Tornado Verification Science from CoreLogic works, and to highlight the unique approach developed to meet the specific needs of catastrophe event responders and insurance claims management. It is important to understand the limitation of other verification methods currently being used to see why such approaches are not the ideal methodology to meet the unique needs for the targeted CoreLogic customer's applications.

Limitations of Verification Methods

In this paper we will explore the following sources of data, and their effectiveness when used for forensic tornado verification:

- Visual Reports
- National Weather Service (NWS) Damage Surveys
- Radar-based Algorithms
- Drones and aerial imagery

Using Visual Reports for Tornado Verification

Visual tornado reports, of either the funnel or damage, are the most common and quickest source of information that a tornado has occurred or is currently occurring. Despite their limitations, they are often used by insurers to alert their catastrophe response teams to expect concentrated areas of impacted policyholders. The NWS collects reports from several volunteer sources, including law enforcement, emergency managers, trained spotters, storm chasers, media outlets, and the general public.

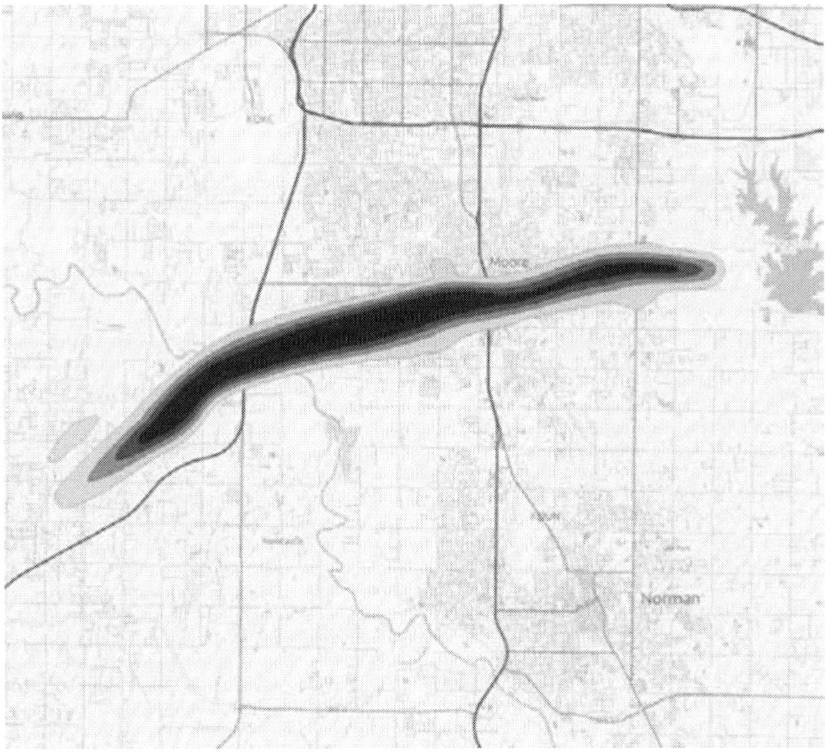

Figure 38.

While these reports are the only publicly available source of ground truth, they have several major limitations. The first, is that initial reports do not provide any information on the intensity, length or width of the tornado, making it impossible to estimate what assets or how many policies could have been impacted. Moreover, visual confirmation is hindered in many cases due to terrain, vegetation, heavy rainfall, and darkness.[152] The second limitation is the ambiguous location of the initial tornado report, which, depending on the description, could be located within the path or at

[152] Blair, S.F. and J.W. Leighton, 2014: Assessing Real-Time Tornado Information Disseminated through NWS Products. *Wea. Forecasting*, 29, 591– 600.

the observer's location outside the path. Finally, most tornadoes are associated with a single tornado report, but some tornadoes are associated with multiple initial reports, complicating the estimate of the number of tornadoes that actually occurred.

Using NWS Damage Surveys

Ground surveys of damage are carried out by NWS personnel for some tornadoes with the goal of establishing the maximum intensity, maximum width, and the beginning and ending points. For a subset of tornadoes, the survey contains sufficient detail to derive shapefile polygons outlining the different levels of damage. These detailed surveys provide the best representation of damage that has occurred on the ground and can be used to estimate the number of impacted homes and assets. Unfortunately, these detailed surveys are only produced on a case-by-case basis depending on the level of impact and the availability of personnel to conduct the survey. Furthermore, these surveys are usually not made available until the day(s) following the event, greatly delaying impact and exposure estimates.

Using Weather Radar for Tornado Verification

Providing the public with sufficient warning lead time for potential tornadoes is a crucial role of the NWS. Although storm spotters provide valuable information on the visual appearance of the storm, these observations are inconsistent and carry the same limitations previously mentioned with tornado reports. Therefore, weather radars are used extensively to find and diagnose storms that are capable of producing a tornado. Weather radars can cover large areas by transmitting energy into the atmosphere and measuring the return echo (reflectivity) and the component of the wind blowing towards or away from the radar (radial velocity). Weather radars are an invaluable tool in consistently estimating precipitation and wind speeds across larger areas of the country that would otherwise be absent of observations.

To this end, several automated algorithms have been developed by the National Severe Storm Laboratory (NSSL) to diagnose the presence and characteristics of rotation in thunderstorms. The first is the Tornado Detection Algorithm (TDA)[153] developed to alert NWS forecasters of developing or ongoing tornadic storms. The algorithm triggers a tornado detection based on wind data at different altitudes and times. Because the algorithm was designed as an aid to forecasters, it maximizes detection at the expense of increased false alarms, depending on the human forecaster to mentally remove false signatures.

Another NSSL product, Rotation Tracks,[154] was developed to deliver a more comprehensive but simplistic view of storm rotation. By tracking the strongest low-level rotation from neighboring weather radars over a specified time interval, coherent swaths of rotation can be highlighted rather than individual tornado detection points. When compared to the detailed NWS damage surveys, experimental tornado paths using the Rotation Tracks product demonstrates large variability in skill and suggests the product is more useful as an aid for manually drawn paths than automated paths.[155] Although several quality control measures were introduced, the algorithm still misidentifies areas of rotation due to anomalous rotation signatures created when thunderstorms move over wind turbine farms and rotation in portions of the thunderstorm that are not capable of producing tornadoes.

Between 2012 and 2013, the NWS enhanced the capability of their radar network by adding dual polarization technology. It works by measuring the difference between horizontal and vertical energy waves that each behave differently depending on the size and shape of the particles in the air. The upgrade allows the radar to identify areas of hail and rain and even differentiate between big drops and small drops.

---

[153] Mitchell, E.D., S.V. Vasiloff, G.J. Stumpf, A. Witt, M.D. Eilts, J.T. Johnson, and K.W. Thomas, 1998: The National Severe Storms Laboratory Tornado Detection Algorithm. *Wea. Forecasting*, 13, 352–366.

[154] Miller, M.L., V. Lakshmanan, and T.M. Smith, 2013: An Automated Method for Depicting Mesocyclone Paths and Intensities. *Wea. Forecasting*, 28, 570 –585.

[155] Karstens, C. D., and Coauthors, 2016: Evaluation of near real-time preliminary tornado damage paths. *J. Oper. Meteor.*, 4, 132 –141.

In the case of tornadoes, debris, such as leaves, dirt, or building material, is often lofted into the air, producing a distinct signal from hail and rain.[156] Unfortunately, an operational tornado debris detection product has not yet been implemented by the NWS.

Using Drones and Aerial Imagery

- Data capture/processing
  - Expensive
  - Time lag
  - Ongoing inclement weather hindering data capture
  - Requires platform/software to view data

The CoreLogic Solution

To combat these challenges with tornado verification, CoreLogic has built a tornado verification model that takes advantage of the best aspects of radar data and reports. Our novel comprehensive approach to using all the available radar data with the ability to use public/social media reports and the expertise of experienced meteorologists consistently provides 250-meter resolution with neighborhood-level detail across most of the country.

The proprietary radar algorithm was designed to preserve the high-resolution of the radar-based estimates while mitigating their limitations. Unlike operational algorithms, the negative influence of wind turbines and other ground contamination has been limited and the algorithm appropriately focuses only on areas of the storm that can produce tornadoes. This is done by developing a proprietary radar wind processing techniques and uniquely taking advantage of the dual polarization data through advanced machine learning techniques.

---

[156] Ryzhkov, A.V., T.J. Schuur, D.W. Burgess, and D.S. Zrnic, 2005: Polarimetric Tornado Detection. *J. Appl. Meteor.*, 44, 557– 570.

Corelogic Tornado Verification Advantages
• Comprehensive processing of wind and tornado debris signatures from radar. • Advanced quality control process that allows public reports of damage and visual appearance to be combined with the radar data. • Damage model trained to estimate actual tornado path rather than the presence of a tornado. • Using advanced machine learning techniques to produce probabilistic tornado damage paths with nationwide coverage in real time.

Finally, when present, valuable information from visual reports can be ingested by the algorithm to further refine the tornado paths. This allows us to include valuable visual reports that would otherwise be excluded, increasing the accuracy of the tornado path.

Summary

The Property & Casualty industry is exposed to billions in insured losses every year due to tornadoes. Knowing the location and scope of the damage from tornadoes allows insures to respond within minutes to hours rather than days.

Verify Tornado Claims to Improve Customer Satisfaction and Profitability

With Tornado Verification Technology from CoreLogic, carriers can compare detailed storm maps to their book of business to better understand the impact of each unique storm, fast-track obvious claims, and allocate appropriate resources toward more difficult or suspicious claims. Because our Tornado Verification provides precise knowledge of when tornadoes likely impacted a specific property, insurance claims adjusters can:

- Objectively handle, document, and communicate tornado claim decisions
- Triage resources to the greatest impacted areas

- Proactively reach out to insureds for emotional goodwill and build stronger brand loyalty
- Close claims faster for improved customer satisfaction
- Correctly identify losses that were potentially caused by pre-policy storms

Figure 39.

A Powerful Combination

Weather Verification Services represent a powerful dataset for property and casualty insurers. With the information provided by these databases, insurance companies can create new solutions and workflows to close claims faster, ultimately increasing customer satisfaction and brand loyalty, while more accurately identifying suspicious claims.

Document Submitted by Representative Randy Weber

Comments of Julienne Sugarek, Vice President of Power Delivery Solutions on Behalf of CenterPoint Energy, Inc., before the United States House of Representatives, Committee on Science, Space, and Technology, Subcommittee on Energy, Hearing Entitled "The Future of Electricity Delivery: Modernizing and Securing our Nation's Electricity Grid," Hearing Held on July 17, 2019, Comments Submitted July 31, 2019

Introduction

CenterPoint Energy appreciates the opportunity to provide comments to the United States House Committee on Science, Space, and Technology, Subcommittee on Energy for their July 17th, 2019 hearing titled "The Future of Electricity Delivery: Modernizing and Securing our Nation's Electricity Grid." CenterPoint Energy supports legislation that would benefit enhanced grid resiliency and emergency response; intelligent grid modeling, visualization, architecture, and controls; a technology demonstration grant program; grid-scale energy storage; hybrid energy systems; grid integration; grid cybersecurity; and protection of sensitive grid information. Due to the nature and geographic location of our business operations, our company size, and our approach to grid modernization, cybersecurity, and emergency preparedness and response, our company can provide a unique perspective and expertise on the topics addressed in the hearing. We would welcome additional discussion on how to advance and modernize the electrical grid to respond to security and natural disaster incidents.

About CenterPoint Energy

Headquartered in Houston, Texas, CenterPoint Energy, Inc. (NYSE: CNP) is an energy delivery company with regulated utility businesses in eight states and a competitive energy businesses footprint in nearly 40 states. Through its electric transmission and distribution, power generation, and natural gas distribution businesses, the company serves more than 7 million metered customers in Arkansas, Indiana, Louisiana, Minnesota, Mississippi, Ohio, Oklahoma, and Texas for natural gas and electric service. Two and a half million of the 7 million customers are served by our electric transmission and distribution businesses in Houston, Texas and Indiana. Within our southern Indiana footprint, we also provide power generation services with a capacity of nearly 1,300 megawatts. CenterPoint Energy's competitive energy businesses include natural gas marketing and energy-related services; energy efficiency, sustainability, and infrastructure modernization solutions; and construction and repair services for pipeline systems, primarily natural gas. The company also owns 53.8 percent of the common units representing limited partner interests in Enable Midstream Partners, LP, a publicly-traded master limited partnership that owns, operates, and develops strategically-located natural gas and crude oil infrastructure assets. With approximately 14,000 employees and nearly $34 billion in assets, CenterPoint Energy and its predecessor companies have been in business for more than 150 years. For more information, visit CenterPointEnergy.com.

Intelligent Grid & Modernization

CenterPoint Energy believes that a resilient and reliable electric grid is critical to serving our customers' needs. Our customers, just like individuals and businesses across the United States, all want electricity that is reliable, resilient, and affordable. We provide power to some of the nation's largest companies whose products are consumed by millions around this country and the world. Our electric operations in Houston serves customers such as the world-renowned Texas Medical Center who cares for more than 10 million patients a year.[157] Ensuring that electricity is flowing is critical to the economic vitality of the nation's fourth-largest city, Houston.

Electric grid infrastructure has become a target for hostile cyber and physical attacks, from nation-states attempting to damage and disrupt electric infrastructure to individuals shooting at electrical substations.

[157] https://www.tmc.edu/about-tmc/facts-and-figures/.

Within our Houston, Texas footprint, CenterPoint Energy has invested in an enhanced multi-component intelligent grid. The grid is comprised of power line monitoring equipment, remote switches, and other automated equipment that will locate and help restore power outages as they occur. The grid also includes more than 2.4 million smart meters across a 5,000-square-mile area, and all the computing systems interface to allow these systems to communicate with one another. Layered on top of this infrastructure are key management and employees who actively monitor and manage these systems to minimize power interruptions to our customers. Smart meters automate meter reading and service connection and disconnection, as well as provide consumers more frequent, detailed information on their electricity use. Our grid performs approximately 3 million electronic service orders annually with a 99.8 percent average success rate. This automation has saved the company nearly 1.9 million gallons of fuel and avoided more than 17,000 tons of C02 emissions. The intelligent grid allows CenterPoint Energy to monitor the health of devices and assists in reducing the number and length of power outages. More than $20 million is saved annually by our customers from connection and disconnection service order fees that have been eliminated as a result of remote metering. It also allows for Retail Electric Providers (REPs) to provide new products and services.

With the deployment of our intelligent grid, we have also deployed a Power Alert Service (PAS). The PAS serves our Houston electric area customers with texts, emails, or phone calls (at the customer's direction). This free tool notifies customers about power interruptions at or near their address and keeps them apprised about estimated restoration times throughout the outage event. Nearly 1.35 million customers are enrolled in PAS. Our intelligent grid has saved over 310 million outage minutes since 2011 for our customers.

CenterPoint Energy is at the forefront of grid modernization innovation. Our electric operations business has several projects slated for 2019 and beyond that are aimed at advancing our electric grid in Houston, Texas and southern Indiana. These initiatives include: 1) Transformer on-line monitoring pilot to understand substation transformer health and predict and prevent transformer failures; 2) Advanced Metering System (AMS) voltage data analytics proof of concept to identify potential failures before they occur; 3) Piloting the ability to remotely retrieve relay event data, which would eliminate the need for field crews to retrieve data manually; 4) Advanced Distribution Management Systems (ADMS) upgrades to advance our ability to incorporate Distributed Energy Resources (DERs); 5) Conducting a proof of concept to allow for remote access to Intelligent Grid Switching Device (IGSD) for data retrieval; 6) Conducting proofs of concept for DER Interconnection protection to optimize existing infrastructure and intelligent grid technology to allow for emerging technologies to connect to our grid; 7) Advanced Metering Infrastructure (AMI) deployment including meter deployment and ADMS integration; 8) Integration of AMI data into our ADMS in our Indiana territory; 9) Conservation Voltage Reductions (CVR) to reduce energy use and system losses; 10) Deployment of solar and storage projects that are interconnected to our grid in Indiana; and 11) The Post House, a building that is currently under construction, which will serve as a research platform to test cutting edge energy technology in a real world setting in downtown Evansville, Indiana.

These projects will support advancements in our electric grid capabilities, address changing customer expectations, diversify our electric generation resources in Indiana, develop more robust communications networks, enhance our distribution systems, enhance our PAS capabilities, enhance our predictive asset and power quality analytics, support electric vehicle adoption, improve our operational efficiency, and realize grid and customer value from our grid technologies. CenterPoint Energy sees great potential in supporting research and development to advance projects such as these to further advance our electrical grid capabilities and leverage advanced technological capabilities, such as artificial intelligence and machine learning. CenterPoint Energy has a dedicated Energy Innovation Center (Center) that showcases innovative educational and technological demonstrations, including battery storage, micro/nano grid energy storage, unmanned aerial systems (drones), connected communities, and artificial intelligence. The Center demonstrates innovative technologies utilized in our electric and natural gas operations. We have hosted more than 1,000 tours for industry leaders, government officials, community groups, and students.

Cybersecurity

Ensuring data, systems, and infrastructure are safe and secure is critical in serving our customers. CenterPoint Energy maintains a state-of-the-art Cyber Security Operations Center that works proactively with our electric operations to protect our grid. Given our geographic footprint, our operations benefit from centralized emergency response, cybersecurity, and physical security operations teams. The centralization of our experts and key management has enabled our company to be engaged in both local and national discussions. We routinely work with local, state, and federal agencies to secure our infrastructure throughout our geographic footprint. In addition, we are engaged in infrastructure discussions with the American Gas Association (AGA), Edison Electric Institute (EEl), Electric Power Research Institute (EPRI) and the National Cyber Security & Communications Integration Center (NCCJC). The Greater Houston Partnership has established a cybersecurity task force that is comprised of 16 critical sectors, as identified by the Department of Homeland Security (DHS), in which CenterPoint Energy has been an active participant. This taskforce has developed cybersecurity best practices that can be leveraged by organizations and companies of all sizes. Our teams have also actively participated in the San Antonio Fusion Center where we have focused on how to further secure our infrastructure.

Electromagnetic Pulse Protection

CenterPoint Energy continues to assess other infrastructure threats such as electromagnetic pulse (EMP) threats as identified by President Trump's recent Executive Order.[158] We have worked with industry leaders and research institutions to develop and validate a cost-effective alternative to address EMP threats in electric utility substations and other parts of the

[158] https://www.whitehouse.gov/presidential-aclions/executive-order-coordinating-national-resilience-electromagnetic-pulses/.

bulk electric system. The result of these efforts has been a scalable turnkey EMP protection alternative for the electric industry, providing a cost-effective means to mitigate EMP effects using proven technologies that can be applied throughout the U.S. With an initial installation already in service on CenterPoint Energy's electric transmission system, we are working to facilitate the commercial expansion of this EMP protection technology for use by other utilities in the U.S.

Hurricane Harvey Response and Restoration

In August, 2017, a Category 4 storm made landfall near Port Aransas, Texas. Hurricane Harvey stalled after initial landfall, impacting Texas and Louisiana for days. The maximum sustained winds were 130 mph at landfall, and 60.58 inches of rain deluged southeast Texas. This amount of rain broke the Texas single-storm record of 48 inches set in 1978 and exceeded the 10-year annual average rainfall amount. More than 42,000 lightning strikes were recorded across our electric service territory. In addition, tornadoes spawned across southeast Texas, Louisiana, Alabama, Mississippi, Tennessee, and North Carolina. Utilizing CenterPoint Energy's robust emergency operations plan, 2,200 employees were activated, and an additional 1,500 contractors and mutual assistance personnel joined to assist in restoration.

During our response to Hurricane Harvey, we operated more than 250 distribution automation devices such as intelligent grid switches, which enhanced service to 140,000 customers. Our crews performed more than 1.27 million total restorations. We saved 41 million outage minutes for our customers through automation on our grid. Our PAS made 352,629 outage notifications to customers. Our intelligent grid also utilizes AMS meters that had 45,000 orders remotely processed with a 97% accuracy rating. During the recovery, we were able to remotely bill 700,000 customers with actual readings 98.9% of the time. The data collected from these meters allows our employees to use real-time analytics to assess, monitor and resolve issues. As part of our restoration efforts, we used 15 drones to perform damage assessments and evaluate working conditions for 500 locations within our footprint. Leveraging this technology allowed our employees to efficiently direct crews to accessible locations. Our crews attached infrared technology to the drones that allowed them to identify equipment that needed further inspection, effectively speeding up our restoration time.

Due to a flooded substation, CenterPoint Energy utilized a 50MVA mobile substation that provided electric service to more than 9,000 customers without power. The mobile substation offered flexibility and reliability that would not have existed had we not been appropriately prepared to stage this mobile substation.

Grid Modernization Research and Development Act and the Grid Cybersecurity Research and Development Act

CenterPoint Energy applauds members of Congress for prioritizing the nation's electric grid. The Grid Modernization Research and Development Act and the Grid Cybersecurity Research and Development Act emphasize the need for additional research and development in securing our nation's electric grid. CenterPoint Energy supports legislation that would benefit enhanced grid resiliency and emergency response; intelligent grid modeling, visualization, architecture, and controls; a technology demonstration grant program; grid-scale energy storage; hybrid energy systems; grid integration; grid cybersecurity; and protection of sensitive grid information.

Cybersecurity is one of the biggest threats to our grid and this threat will continue to grow as we enhance and expand our grid. The Electric Power Research Institute (EPRI) and Edison Electric Institute (EEI) are leading efforts in cybersecurity and we recommend that Congress support these types of efforts to advance our grid. Further opportunities for Congress to explore would include addressing research for recovery after successful cyber-attacks. Funding could be directed towards researching ways to accelerate restoration and recovery from attacks that occur. In addition, Congress should support the development of language that permits the sharing of cybersecurity information about real or perceived threats. Currently, utility professionals may limit sharing vital information for perceived violation of North American Electric Reliability Corporation

(NERC) or other relevant regulations. For the utility sector to be agile and effective in mitigating cybersecurity concerns, utilities need to be able to have a free flow of information about risks with other utilities while minimizing risk of disclosure of critical and/or sensitive information to the public.

Our electric operations footprint has seen increases in distributed energy resources (DER) deployment over the last 10 years and we expect that to continue. As of April 2019, we have connected more than 6,000 DERs in the Houston area and more than 490 DERs in Indiana. Congress should act to support DER installations that are installed in a way that maintains the integrity and resiliency of the grid. Our Houston operations is a Transmission and Distribution Service Provider (TDSP) and we are not able to own generation assets, but we do need to be able to support customers who wish to connect generation assets to our system.

For utilities that operate in deregulated markets as TDSPs, such as our Houston, Texas operations, we would like to have the opportunity to be able to propose storage applications that contribute to grid stabilization (reliability) only. It would be beneficial for us to be able to study the impact of storage technologies and benefits they may offer to the grid - such as voltage regulation -so we can better understand the potential value for our customers. If Congress can support research and development in storage applications this would improve our grid security, resiliency, and reliability. In addition, this would allow us to better understand how we can provide better services to our customers.

Conclusion

CenterPoint Energy thanks the Subcommittee members for allowing comments to be submitted and for demonstrating strong leadership in advancing modernization initiatives that support reliability and resiliency of the grid. We welcome the opportunity to provide additional input on how to protect and advance the nation's electric grid.

Charts Submitted by Representative Francis Rooney

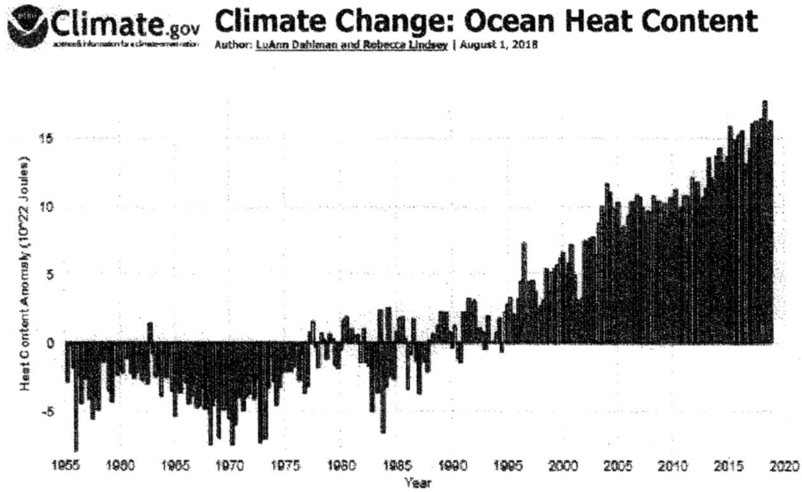

Figure 40. Explore this interactive graph: Click and drag to display different parts of the graph. To squeeze or stretch the graph in either direction, hold your Shift key down, then click and drag. This graph (source data) shows differences from the long-term average global ocean heat content (1955-2006) in the top 700 meters of the ocean.

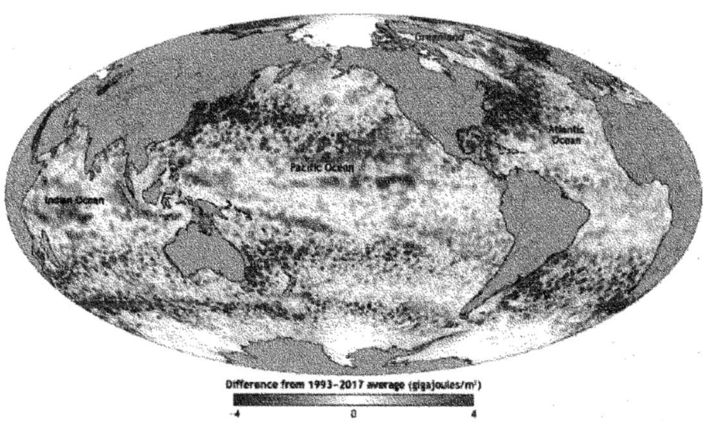

Global map showing where 2017 heat content in the top 700 meters (2,300 feet) of the ocean was higher (orange) or lower (blue) than the 1993–2017 average. NOAA Climate.gov map, adapted from *State of the Climate in 2017.*

Figure 41.

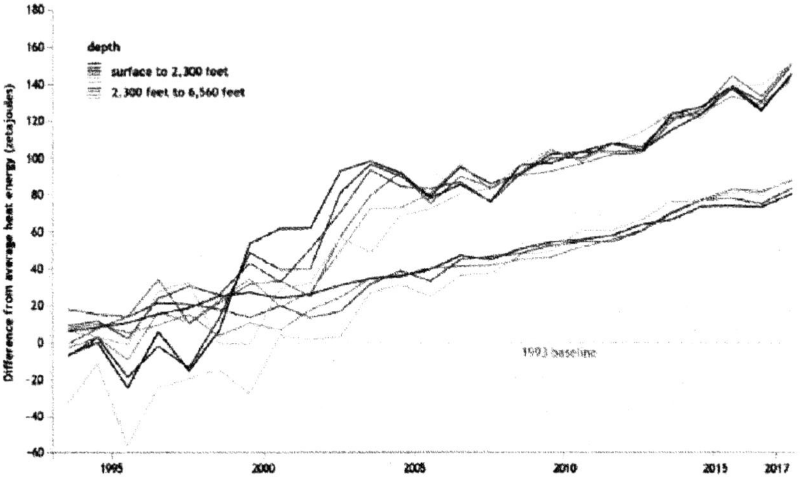

Ocean heat content over time from the sea surface to a depth of 700 meters (2,300 feet) in shades of orange, and from 700-2,000 meters (6,600 feet) in shades of purple. (Different colors represent different data sets. Details can be found in Figure 3.6 of *State of the Climate in 2017*.) The upper ocean and deeper ocean trends are both strongly positive, with 2017 exhibiting record-high values for both. Graph by NOAA Climate.gov, adapted from *State of the Climate in 2017*.

Figure 42.

In: Climate Change
Editor: Nicolas Delgado

ISBN: 978-1-53619-113-4

Chapter 2

CREATING A CLIMATE RESILIENT AMERICA: REDUCING RISKS AND COSTS*

Select Committee on the Climate Crisis

Wednesday, November 20, 2019
House Of Representatives, Select Committee On The Climate Crisis,
Washington, DC.

The committee met, pursuant to call, at 1:32 p.m., in Room 1334, Longworth House Office Building, Hon. Kathy Castor [chairwoman of the committee] presiding.

Present: Representatives Castor, Bonamici, Levin, Casten, Graves, Carter, and Miller.

Ms. CASTOR. The committee will come to order.

Without objection, the chair is authorized to declare a recess of the committee at any time. You know, decisions are made every day about where and how to build, what home to buy, and how to prepare for the

---

* This is an edited, reformatted and augmented version of Hearing before the Select Committee on the Climate Crisis, House of Representatives, One Hundred Sixteenth Congress, First Session, Serial No. 116–15, dated November 20, 2019.

impacts and costs of the climate crisis, and, today, we will talk about the kinds of climate risk information standards and tools that communities need to reduce the risks and costs of climate change, including more extreme floods and wildfires.

So I will recognize myself for 5 minutes for an opening statement. Last week, the committee heard from members directly about the impacts of the climate crisis on the communities they represent all across America, and today we will examine the Federal Government's role in helping communities reduce the risks and costs of climate fueled disasters. States, cities, and Tribes across the country are taking bold action to adapt to climate change. They need a strong Federal partner, whether it is through scientific information or technical assistance tools, the Federal Government has an opportunity to help communities grow stronger in the face of the climate crisis with a particular eye to communities that are on the front lines.

Every community is different, which is why the best role for the Federal Government is to empower local communities to give them the right tools and data to build strong and to rebuild smart. The Federal Government also can lead by example by requiring that federally funded projects avoid areas that are prone to the worsening effects of the climate crisis. One thing we can't do is move backwards.

In 2017, President Trump decided to rollback a Federal flood standard meant to protect communities from damage. Ten days later, the necessity of those protections was made evident when Hurricane Harvey struck Texas in the Gulf Coast causing massive loss of life and property. Across the Nation, more than 20 states and hundreds of communities have adopted higher standards to reduce flood losses through establishing higher elevation requirements or limiting development in flood-prone areas.

Now the Federal Government must modernize the National Federal Flood Standards to ensure the resilience of federally supported development, redevelopment, and rebuilding. To help local decisionmakers better protect their citizens, we must also make sure they know how climate change is increasing risks in their communities. Whether it is flooding, wildfires, or extreme heat, they need to know what to anticipate as well as the best ways to prepare their residents for a changing climate.

We must also establish clear, uniform national standards that are grounded in robust climate science. With better guidance, local officials can make better decisions about where to build homes, schools, and hospitals. Maps that integrate climate risk will help us make better decisions today so our buildings can meet the demands of the future. As more and more climate risk data becomes available, the Federal Government will need to develop new maps to take into account the increasing effects of the climate crisis, including future sea level rise and stronger storms.

We also need to better understand how the risk of wildfires threatens communities, forests, and Federal assets. Experts across the country are working on rebuilding and landscape designs to help families and communities respond to the growing risk of wildfires, but that research doesn't always prompt better choices at least at the scale needed to reduce wildfire losses. We can change that. We can enhance our current maps, codes, and standards in order to protect Americans against floods and wildfires, we can equip local governments with the tools they need to build resilient infrastructure, and we can make sure the Federal Government leads by example.

We need to act quickly. Natural disasters in the United States have become more frequent and more severe and more costly over the past two decades. In fact, since the year 2000, flood-related disasters in the United States caused more than $845 billion in losses, making it the costliest disaster threat in the Nation, and during the last three hurricane and wildfire seasons, our country experienced $330 billion in damages from six hurricanes as well as over $40 billion from eight wildfires.

But here is the good news: Through serious climate action, we can reduce these costs. Resilient communities attract investments, reduce dependence on Federal disaster aid, and protect their public credit ratings, which can reduce the cost of capital. Resilient communities can make sure land use decisions avoid flood and wildfireprone areas. They can adopt and enforce good codes and standards. They can make sure residents are well prepared for storms and wildfire seasons. They just need our help to get there.

The solutions we discuss today will uplift communities, protect valuable natural resources, and reduce the cost of the climate crisis. I look forward to hearing from this great panel of experts to help guide our decisionmaking. And at this time, I will recognize the ranking member, Mr. Graves, for 5 minutes.

[The statement of Ms. Castor follows:]

OPENING STATEMENT (AS PREPARED FOR DELIVERY) REP. KATHY CASTOR (D–FL), CHAIR U.S. HOUSE SELECT COMMITTEE ON THE CLIMATE CRISIS "CREATING A CLIMATE RESILIENT AMERICA: REDUCING RISKS AND COSTS" NOVEMBER 20, 2019

Last week we heard from Members directly about the impacts of the climate crisis on the communities they represent all across America. Today we'll examine the federal government's role in helping communities reduce the risks and costs of climatefueled disasters.

States, cities and tribes across the country are taking bold action to adapt to the changing climate. They need a strong federal partner. Whether it's through scientific information on what the future holds, climate risk data, resilience standards or technical assistance tools, the federal government has an opportunity to help communities grow stronger in the face of the climate crisis—with a particular eye to communities that are on the front lines.

Every community is different. Which is why the best role for the federal government is to empower local communities, to give them the right tools and data to build strong and rebuild smart.

The federal government also can lead by example, by requiring that federallyfunded projects avoid areas that are prone to the worsening effects of the climate crisis.

One thing we can't do is move backwards. In 2017, President Trump decided to roll back a federal flood standard meant to protect communities from damage. Ten days later, the necessity of those protections was made evident when Hurricane Harvey struck the Texas Gulf Coast, causing massive loss of life and property.

Across the nation, more than 20 states and hundreds of communities have adopted higher standards to reduce flood losses, through establishing higher elevation requirements or limiting development in flood-prone areas. Now the federal government must modernize the national federal flood standards to ensure the resilience of federally-supported development, redevelopment, and rebuilding.

To help local decision-makers better protect their citizens, we must also make sure they know how climate change is increasing risks in their communities. Whether it's flooding, wildfires or extreme heat, they need to know what to anticipate, as well as the best ways to prepare their residents for a changing climate.

We must also establish clear, uniform national standards that are grounded in robust climate science. With better guidance, local officials can make better decisions about where to build homes, schools and hospitals.

Maps that integrate climate risk will help us make better decisions today, so our buildings can meet the demands of the future. As more and more climate risk data becomes available, the federal government will need to develop new maps that take into account the increasing effects of the climate crisis, including future sea-level rise and stronger storms.

We also need to better understand how the risk of wildfires threatens communities, forests, and federal assets. Experts across the country are working on building and landscape designs to help families and communities respond to the growing risk of wildfires. But that research doesn't always prompt better choices, at least at the scale needed to reduce wildfire losses.

We can change that. We can enhance our current maps, codes and standards in order to protect Americans against floods and wildfires. We can equip local governments with the tools they need to build resilient infrastructure. And we can make sure the federal government leads by example.

We need to act quickly. Natural disasters in the United States have become more frequent, more severe, and more costly over the past two decades. In fact, since 2000, flood-related disasters in the United States caused more than $845 billion in losses, making it the costliest disaster threat in the nation. During the last three hurricane and wildfire seasons, our country experienced $330 billion dollars in damages from six hurricanes, as well as over $40 billion from eight wildfires.

Here's the good news: through serious climate action, we can reduce these costs across the nation. Resilient communities attract investments, reduce dependence on federal disaster aid, and protect their public credit ratings, which can reduce the cost of capital. Resilient communities can make sure land-use decisions avoid flood and wildfire-prone areas. They can adopt and enforce good codes and standards. They can make sure residents are well-prepared for storms and wildfire seasons. They just need our help to get there.

The solutions we discuss today will uplift communities, protect valuable natural resources, and reduce the costs of the climate crisis. I look forward to hearing from our great panel of experts.

Mr. GRAVES. Thank you, Madam Chair. I want to thank you all for being here today. I am very much looking forward to your testimony. We have votes going on right now in the Transportation Committee where I am supposed to be, and I have a bill on the floor in a little while so I am going to be bouncing around a bit, but this topic, I think, while the chair and I don't see eye to eye on everything, almost everything, though, right? This is an area where we do have very strong, I think, consensus among the members of the committee. No matter what we do with emissions, we are going to continue to see our seas rise and our coastal communities being more vulnerable. As we have discussed in this committee before, you can look around at the coastal counties, parishes, and boroughs around the

United States and those only constitute about 10 percent of the land area of this country, yet over 40 percent of the population lives there.

With sea rise, as the chair noted, increasing disaster response and recovery cost, it is not an option, I think, with that percentage of the population living there and the number going up, for us to just say, well, you all are going to have to move or there is nothing we can do. The reality in my opinion is that we have got to get good at resilient living. We have got to get good at doing that. We also need to make sure that we are being very thoughtful about new development. A recent analysis by the Congressional Budget Office found that the majority of increased disaster costs were actually attributable to development in those areas, meaning we are developing in areas that are vulnerable and not being thoughtful about how to do it.

Now, President Obama's executive order related to floodplain rebuilding and standards was mentioned by the chair earlier, and I have to say that I actually did not support that and I did support the withdraw, but let me clear: I supported the intention of it, but I think it was the right move to withdraw and think a little bit more about how to properly do it because of this. The folks that I represent in south Louisiana, we drain two-thirds of the United States, two-thirds. When folks send us more water, we become more flood-prone or more vulnerable. Hurricanes are exacerbated, the impact are exacerbated by the 2000 square miles of coastal land loss we have experienced, not because of anything we did, because of what the Corps of Engineers did.

So, if you were to apply that executive order, you had some parishes in Louisiana, where I think they said 80 percent of the parish you couldn't live there. That is not right. What you have to do— and Administrator Fugate and I have had this discussion. You have got to integrate an offense and a defense strategy into the Flood Insurance Program. For example, let's look at the fact we have $100 billion backlog in Corps of Engineer projects, many of which are designed to address resiliency.

How are we out there talking about telling communities they can't live in certain areas whenever there is a Corps of Engineers' authorization that has been sitting out there in the books for decades, and we can't figure out how to move forward on it? How are we going to move in a direction of resiliency when we made so much progress last year with the Disaster Recovery Reform Act, with all of the funds through the Community Development Block Grant Disaster Recovery Flood Mitigation program through the Corps of Engineers, construction general program last year record funds and then move into this year whenever we are doing a Community Development Block Grant Disaster Recovery bill—just this week, we put prohibitions in there and say that you can't use the funds on authorized Corps of Engineer projects that actually make your community more resilient. It doesn't make sense. It doesn't. And, obviously, not thoughtful policy has been progressed by this House within the last few months.

This is an area where we should all be working together. This isn't a fight over who is right, who is wrong. We all agree. We all agree that our coastal communities are more vulnerable. We all agree that adaptation measures make sense. We all agree in bipartisan support for the legislation last year, for the appropriations last year, bipartisan support, but we have got to continue building upon that progress because the district that the chair represents, the district that I represent, the district that Mr. Levin represents, we represent coastal areas, coastal communities, people that are vulnerable, and you will not find a divide among us in terms of us wanting to make sure—do you have a coast?

Mr. CASTEN. No.

Mr. GRAVES. I didn't think so. A little coastal envy over there. I see it. You come vacation in our districts. It is fine. You can keep coming. But we need to make sure that we are being thoughtful and that all the policies of this Congress are continuing to move— and I think this committee is a great place to do it—continuing to move in this direction of ensuring that we can have resilient communities, resilient ecosystem, but that we have that offense and defense at the table and aren't just coming in and drawing areas and say "You can't live here." I think that there is a smart way of doing it. I think, in some areas, we are going to have to tell people that you can't populate these areas, and we have done that in South Louisiana, but I think that we have got to be very thoughtful using all the tools in the toolbox as we move forward. Looking forward to you all's testimony.

And I yield back the time that I don't have. Ms. CASTOR. Thank you, Mr. Graves.

Without objection, members who wish to enter opening statements into the record may have 5 business days to do so.

Now, I want to welcome our witnesses.

Welcome, Craig Fugate. He was the administrator of the Federal Emergency Management Agency for nearly 8 years in the Obama administration and also served as the Florida Emergency Management Director under Governor Jeb Bush. He is more popular in Florida than a rock star. Maybe not Jimmy Buffett, but he currently provides senior level advice and consulting in the area of disaster management and resiliency policy.

Alice Hill is the Senior Fellow for Climate Change Policy at the Council on Foreign Relations. Judge Hill's work focuses on responding to the risks and consequences of the climate crisis. Prior to joining the Council on Foreign Relations, she served as Special Assistant to President Obama and developed national solutions as Senior Director for resilience policy for the National Security Council staff.

Chad Berginnis is the Executive Director for the Association of State Floodplain Managers and previously worked in the Ohio floodplain management program and was Ohio's State hazard mitigation officer. He is recognized as an expert in floodplain management with more than 25 years of experience in natural hazard management, flood loss reduction, and land-use planning.

Without objection, the witnesses written statements will be made part of the record.

With that, Mr. Fugate, you are now recognized to give a 5minute presentation of your testimony.

Statements of the Honorable W. Craig Fugate, Craig Fugate Consulting LLC, Former FEMA Administrator; the Honorable Alice Hill, Senior Fellow for Climate Change Policy, Council on Foreign Relations; and Chad Berginnis, Executive Director, Association of State Floodplain Managers

Statement of the Honorable W. Craig Fugate

Mr. FUGATE. Well, thank you, Chair Castor and Ranking Member Graves, and the other members of the committee. First of all, I need to acknowledge that, you know, when you leave Federal Government and you are by yourself, you don't have much of a support system, so I did enlist some help in getting ready for this hearing. Lars Anderson, Blue Dot Strategics, I worked with very closely, and we worked with The Pew Charitable Trusts on flood policy. So, full disclosure, I do work for those folks, but my space and area that I really have focused on is asking the question, why are disaster costs going up exponentially?

We are not necessarily seeing more disasters, but the impacts of the disasters have been climbing to the point where the General Accounting Office has now put climate change impacts, particularly the cost of disaster, on the high risk list. If we talk about the $850 billion, which are both direct and indirect, the General Accounting Office points out, since 2005, almost half a trillion dollars in direct Federal funding has gone to disaster response. And I like to remind people that those costs are uninsured losses. FEMA does not pay for insured losses. They only pay for uninsured losses. And as this committee wrestles with how do we build resilience, I think we need to answer a fundamental question: Why did the Stafford Act see a disincentive in maintaining insurance in localand government-owned buildings at the state and transfer that risk to the taxpayer? Think about it. Every time you see these big FEMA disaster dollars going out the door, it is either because families didn't have insurance or were underinsured or and the big dollars are the massive amounts of money that go out because local and State governments are self-insured, which is really not insurance. It is a game they play that says, if it is really bad, we hope somebody else will bail us out at 75 percent on the loss.

And because FEMA's program under the Stafford Act goes back to the first dollar, there has been no virtual increase in state and local spending on reducing these impacts. And as long as we continue to set the threshold so low on disasters, we are encouraging States not to take action and we are subsidizing the development in high-risk areas, as Representative Graves says. You know, South Carolina sees one of the fastest growing areas in flood-prone areas that can only occur because we subsidize it through the National Flood Insurance Program and through our disaster programs. Just look in your time here the appropriations that were required for both FEMA and HUD, the Corps and others to respond to disasters. Think about what we could be doing with that other money or the fact that we have not had to borrow that money and grown the deficit? And so we need to look at not only policy here, but also I think we need to bring the private sector back into this, and we need to start pricing risk. There is a lot of things we got to do with the Flood Insurance Program and FEMA's trying with Flood Risk 2.0. I think it is a good step. I also caution that it needs to be means-tested because we have a lot of people who don't live on the coast, many in rural parts of the country, agricultural based, and if we go to full actuary value without providing an index of affordability, we are going to price people out of their homes, which I do not think is the intention of Congress. But those that can should be paying what it costs, but I have a simpler solution: Why don't we just stop writing flood insurance for new construction? Let's take care of the folks we got because that program, as much as you have dealt with reauthorization in the short-term program to get that going, it is going to be hard to do any serious reform. There are so many issues we are not going to get to, so let's just take a first step. Let's quit making it worse. If you want to build in a high-risk flood zone, then buy it from the private sector. And it may be a sign that the private sector won't make insurance available or affordable, you shouldn't be building there the way you want to build. And I want to come back to our Federal floodplain management standard.

In 2012, as we were getting ready to go to New Jersey for the infamous non-hug with Governor Christie with President Obama, he turned to me and said: Craig, the debate about climate change is over. We got to start talking about adaptation.

Our problem was we had no tools to say what these impacts looked like and then go, well, what should we be building to? So the Federal flood plain management standard was a very simple idea. It wasn't that you couldn't build in a floodplain; it just said double the amount of height you had to have in the first place. If you were in the National Flood Insurance program, you have to build one foot above base flood elevation. It never said you can't build in a Zone A. All we were saying that if you were taking Federal dollars and building infrastructure with Federal dollars, you had to go to 2 feet, a doubling of that, and if it was a critical facility, like maybe a hospital, a jail, or a 911 center, my first question is, why are you building it in a flood zone in the first place? Build it 3 feet above, and we would provide that additional funding. It did cost more money, but think about the savings we would get. It has been repealed. They are studying it. They have been studying it now since 2017. That tells me we are probably not anywhere close to doing it, and I got bad news for them. I really think we undershot the whole estimate. We should have been talking maybe 3 to 4 feet if you are going to build in a flood zone, and we should have basically said: Stop building in the flood zone, and we need to increase this outside of the 100-year flood zone, which turns out to be not a 100-year flood event anyway. In Houston, they are thinking that may be a 40-year event, and we need to look at what the flood-prone areas are and raise our standards.

The last thing, as I am over, we have a lot of tools that are forecasting the types of climate impacts we are going to have, yet we have not been able to translate that down to decision support tools to local and state governments to go: Well, if you are talking about sea level rise, what does that mean to my community? What does it mean to what I am preparing to do? And so I think these are areas that, again, I have been looking at this problem is, there is no debate about what climate change is being caused by. The only debate is can we adapt fast enough.

Thank you, Madam Chair.

[The statement of Mr. Fugate follows:]

TESTIMONY OF THE HON. W. CRAIG FUGATE, PRINCIPAL, CRAIG FUGATE CONSULTING LLC; FORMER FEMA ADMINISTRATOR, BEFORE THE U.S. HOUSE OF REPRESENTATIVES, SELECT COMMITTEE ON THE CLIMATE CRISIS, "CREATING A CLIMATE RESILIENT AMERICA: REDUCING RISKS AND COSTS" NOVEMBER 20, 2019

Chair Castor, Ranking Member Graves, and Members of the Select Committee, thank you for inviting me to testify today about climate resiliency, and what steps need to be taken to reduce the risks and costs of making our communities more resilient.

As many of you know, I do not mince words when it comes to this topic: The climate has changed and we are seeing more climate driven extreme weather events. It is not something that is 30 years down the road. As a result, we need to start talking about adaptation. Time has run out for debate, action is required.

The stark financial reality today is that the federal government spends billions of dollars annually to deal with the effects of climate change and extreme weather while not spending nearly enough to combat future risk. It is critical that we build in funds for resilience on the front-end of these federal investments. There is a huge cost-benefit to the taxpayer, and the outcome is that disaster relief spending should ultimately be reduced in the out years because it costs significantly less to fund recovery for resilient construction following a disaster.

Disaster Cost Are Growing at an Unsustainable Rate

From the GAO High Risk Report for 2019, Limiting the Federal Government's Fiscal Exposure by Better Managing Climate Change Risks. "Since 2005, federal funding for disaster assistance is approaching half a trillion dollars (about $430 billion), most recently for catastrophic hurricanes, flooding, wildfires, and other losses in 2017 and 2018."

Climate Change Impacts Are Occurring and Getting Worse

Some Highlights from the Findings of the U.S. Global Change Research Program Climate Science Special Report 2017

- Heavy precipitation events in most parts of the United States have increased in both intensity and frequency since 1901 (high confidence). There are important regional differences in trends, with the largest increases occurring in the northeastern United States (high confidence).
- Extreme temperatures in the contiguous United States are projected to increase even more than average temperatures (very high confidence).
- The incidence of large forest fires in the western United States and Alaska has increased since the early 1980s (high confidence) and is projected to further increase in those regions as the climate warms, with profound changes to certain ecosystems (medium confidence).
- Global mean sea level (GMSL) has risen by about 7–8 inches (about 16– 21 cm) since 1900, with about 3 of those inches (about 7 cm) occurring since 1993 (very high confidence).
- As sea levels have risen, the number of tidal floods each year that cause minor impacts (also called ''nuisance floods'') have increased 5to 10-fold since the 1960s in several U.S. coastal cities (very high confidence). Rates of increase are accelerating in over 25 Atlantic and Gulf Coast cities (very high confidence). Tidal flooding will continue increasing in depth, frequency, and extent this century (very high confidence).

The Past Is Not Preparing Us for the Future

As much as we need to learn from past disasters, the topline lesson that needs to be understood is that *we must build, and rebuild after a disaster, for our future risk.* In these scenarios, the past isn't the best indicator of what these risks have been. Many of you all have seen this, unfortunately, in your home districts: we build something back, and it ends up getting destroyed again. We ought to do it differently, and we need to do it better.

Recommendations

Build Better Climate Impact Models and Analysis Tools for States and Local Governments

I have often said that I am not going to debate the merits of climate change. And I saw the effects of it through disaster responses I oversaw at FEMA. It is critical that the risks and effects of climate change are identified and understood so that we can take immediate action.

To that end, the Committee should think about how to accelerate more scientific data and recommendations from a broad cross section of technical and scientific experts, and to consider the need for additional resources to support and improve platforms and models that can forecast and/or characterize sea level rise, flooding probabilities, wildfire risk, drought impacts, and other vulnerabilities associated with extreme weather and changing precipitation patterns.

The Nation Lacks Uniformed Tools to Measure Resilience

As a first step in creating a resiliency standard, develop tools to measure a community's resilience of its Tax Base to natural hazards.

When local officials try to measure resilience, they often talk about critical infrastructure (Power, Water, Communications, etc.). I think a better measure is the resiliency of their tax base to natural hazard risks. From Hurricane Andrew (and the closing of Homestead USAF Base), Hurricane Katrina and the Gulf Coast, Hurricane Michael and the Florida Panhandle, The Camp Fire in California (Paradise), all have seen reductions in their tax base making recovery difficult or delayed. Loss of housing, jobs, and businesses compound the impacts of the disaster and can mean a failure or long delay to recovery.

How and Where Will We Build Matters

Building codes and land use planning are key steps in building resilient communities. Florida has seen the effects of its building codes reducing storm damage. California's 2008 updates to its building codes for wildfire mitigation contributed to homes surviving wildfires in 2017 and 2018. Organizations such as the Institute of Building and Home Safety's Fortified Home program show how building over minimum code requirements can save homes from multiple hazards https://disastersafety.org. Congress should continue to support research in developing model building codes that address climate risk.

Preparing for Extreme Flood Risk

Since leaving FEMA, I have been working with the Pew Charitable Trusts' FloodPrepared Communities initiative on these very issues. Our work aims to decrease the impact of flood-related disasters through cost reduction policies.

I use this as an example of how we need to shift our thinking, investments, and actions as flooding is our nation's most costly natural disaster and affects all 50 states—in areas both inland and coastal. It is something that is impacting constituents in each of your districts and home states. According to the National Oceanic and Atmospheric Administration, flood and coastal storm events have caused, since 2000, nearly $850 billion in overall losses when accounting for impacts such as business interruptions, physical damage to buildings, agricultural losses, and damage to public infrastructure.[1]

Resilience and adaptation are essential to lowering the costs to taxpayers and the risks to our communities. Congress is extraordinarily generous in funding disasters each year to ensure that our communities can

---

[1] National Oceanic and Atmospheric Administration, *Billion-Dollar Weather and Climate Disasters: Summary Stats,* National Centers for Environmental Information, (accessed October 1, 2019) *available at* https://www.ncdc.noaa.gov/billions/summary-stats (considering tropical cyclone to be flood-related disasters).

recover. However, the challenge in this is the inherent bias towards post—disaster assistance over adaptation and pre—disaster mitigation.

It is essential that the federal government alter the long-existing bias that favors post-disaster assistance over federal support for adaptation and pre-disaster mitigation. Investing in resilience is not only good policy that leads to better protection for people and infrastructure, it is a *better* investment in terms of actual dollars. According to one study by the National Institute of Building Sciences, investing in mitigation saves society $6 for each $1 invested.[2]

Congress needs to look at not only the amount of funding for mitigation, but also the types of funding vehicles available. The mitigation needs in one part of the country are different from another, but the underlying commonality is that both the amount of funding and the type of funding is lacking across the board.

We all understand the difficulty in assessing the costs associated with investing in mitigation. This Committee, in particular, and Congress in general, should consider how we are currently looking at mitigation and adaptation costs. Currently, nearly 90 percent of funding for flood risk reduction comes in the aftermath of a big flood. (This is true for most disasters, with the passage of the Disaster Recovery Reform Act of 2018 (DRRA) Federal Fire Management Assistance Grants now included post event mitigation dollars). Obviously, that it is a good thing to rebuild the right way, but we also have to prepare *before* disasters because those investments will be more effective and well-thought out. I would encourage you all to look at how the current analytical approaches may not fully account for the benefits of adaptation and pre-disaster mitigation.

The built environment is, of course, critical to our lives and well-being. However, we must also look at how non-structural solutions can also support adaptation and mitigation efforts in our country. Various nature-based solutions, such as wetlands and parks, can provide self-sustaining flood defenses that support ecosystem restoration while providing

---

[2] https://www.nibs.org/page/mitigationsaves.

recreational space for communities. These have been proven to be across the board 'wins.'

One way the federal government has helped communities create or restore natural open space within floodplains is through FEMA's Pre-Disaster Mitigation (PDM) Program. Through the PDM Program, FEMA has invested in the acquisition of disaster prone or damaged properties with the goal of moving people out of harm's way while creating permanent open space in the process. In theory, this program is a good tool that states, and communities can use to prepare beforehand, but it just doesn't get the funding to make enough of a difference. This needs to change, and I hope that as the Select Committee considers its recommendations, that it encourages increased funding for this at FEMA, as well as support other federal agencies in their evaluation and use of non—structural infrastructure wherever feasible.

A second way for Congress to support resilience is pass the State Flood Mitigation Revolving Loan Fund Act of 2019 (H.R. 1610).

Update Flood Risk Maps and Communication of Flood Risk

I recommend that Congress provide funding to update flood maps to portray all the areas at risk of flooding. For example, many of the homes that flooded in Hurricane Harvey were outside zones where flood insurance was required, which understandably caught homeowners by surprise. The worst thing we can do is create a false sense of security for homeowners and communities. Under the current structure, that's exactly what is happening.

Terms such as a 100-year flood and flood insurance rate maps have led too many to underestimate their flood risk. How to communicate flood risk is terms that home owners will understand can lead to more purchasing flood insurance outside of the Special Flood Risk Areas.

Congress should also require that to participate in the NFIP, the National Flood Insurance Program, states adopt flood hazard disclosure requirements for home sales that provide home buyers a right to know about flood history and risk before going to closure. Currently 29 states have some form of flood risk/history disclosure, 21 states have no requirements.

A key step for homeowners to be resilient is the purchase of flood insurance, either from the NFIP, or from private flood insurers. This action can be taken now by the public, as a first step in developing financial resilience in the face of more extreme flood events.

The Federal Government Should Not Be the First Financial Responder to Frequent Disasters

I would also encourage the Committee to look at how the Federal government response can act as a disincentive for state and local leadership on mitigation and adaptation. The federal government has multiple authorities for providing disaster response and recovery with programs housed in various agencies across the government. For example, direct grants to repair and rebuild public facilities, loans to businesses, families, and local governments, unemployment assistance, special tax treatment of losses, and financial aid to affected individuals all support our communities. A significant portion of this assistance flows through the Disaster Relief Fund (DRF) to function as a complement to state and local resources when disasters overwhelm local and state capacities.

This is critical support that should not be discounted in any way. However, the downside is the increasing number of disaster declaration requests and growing reliance on the federal government. We see this as problematic, not only in terms of federal spending, but also in creating a strong disincentive for local and state leadership on adaptation.[3] *I would encourage the Committee look at the proposals for a "disaster deductible" that FEMA released in 2016 and 2017.*

Stop Growing the Risk

Strengthen requirements for local and state governments, as well as eligible nonprofits, to insure their risk. Too many claim to be self-insured,

---

[3] A 2015 review of state budgeting for disaster concludes that natural disasters and emergencies have not had a significant effect on state finances, "... because states relied on the federal government to provide most of the funding for recovery." https://www.gao.gov/assets/670/669277.pdf.

but have instead transferred their risk to the federal taxpayer when disaster strikes. When the President declares a Federal Disaster under the Stafford Act. No less than 75% of their eligible uninsured losses are required to be covered. This has been an unintended consequence of the Stafford Act, growing the uninsured risk of state and local governments.

One final point I would like to make is about NFIP, the National Flood Insurance Program. As you know, I oversaw this program when I served as FEMA's administrator, and that program has faced a lot of criticism. I am not here to debate the merits of NFIP, as it certainly plays a role in the immediate term to insure existing properties that aren't otherwise insurable. However, when discussing resiliency and mitigation, part of that conversation must include a discussion about *not providing NFIP coverage to new construction in flood zones* that only grows the risk. The question I ask, if the private sector will not insure the risk of new construction is flood prone areas, why should the taxpayer?

Thank you for the opportunity to be here today, and I look forward to answering any of your questions.

Ms. CASTOR. Thank you very much. Judge Hill, you are recognized for 5 minutes.

Statement of the Honorable Alice Hill

Ms. HILL. Thank you very much, Chairwoman Castor and Ranking Member Graves and members of the committee. I am delighted to have a chance to speak with you today. As we have discussed, natural disasters are on the rise and according to the fourth national climate assessment, they will continue to rise as climate change occurs. The costs of climate-driven events is also rising. We have heard about the staggering figures between 1980 and 2018, the National Oceanic and Atmospheric Administration estimates that the costs have been $1.7 trillion. Now, when communities suffer this type of devastation, Americans are generous. The Congress has increasingly authorized supplemental appropriations to provide relief to local and State governments overwhelmed by disasters, to small businesses and individuals who have suffered losses, and to repair damaged Federal assets.

Some of the damage that is wrought by these climate-driven extremes stems from decisions made about how and where people build. Those decisions rest almost entirely with the states and often with local governments. That means that, even though the Federal Government currently has comparatively little say in state and local choices about land use and construction quality, it frequently picks up the tab for those choices after disaster strikes.

The Federal Government's growing generosity to victims of disaster creates a moral hazard. Communities and people place themselves at greater risk because the Federal Government, the Federal taxpayers will bail them out. In the face of accelerating climate change, the Federal Government must reduce the incentives for people to settle in at-risk areas and to build in risky ways. We have heard a lot about flood here today, but let's talk about California's recent experience with wildfire as well. Ten of the most destructive fires in the State of California have occurred since 2015, and the state is currently fighting a vicious wildfire season. Preemptive power shutdowns were occurring in California today. California holds the dubious record for having more buildings destroyed by wildfire than all other States combined. A recent study has estimated that it has more than 2.7 million people and 1.1 million homes located in areas already determined to be at very high risk of fire without looking at the added risk from climate change.

In 2008, California enacted a strict new building code designed to reduce fire risk. During the devastating 2018 Camp Fire, only 18 percent of the 21,100 homes destroyed that were built to older versions of the code survived the fire, but of those houses that were built to the new stricter code, they performed better, but only 50 percent survived. With only half of the homes built to the latest code withstanding climate-fueled wildfires, California cannot assume that its building codes will keep people and property safe. California faces worsening wildfire risk. Its own climate assessment estimates that climate change will expand the burned areas 77 percent by 2100, yet just days after the Camp Fire, Los Angeles County board of supervisors approved a 19,000 home development in an area that the State had already determined is at high or very high risk without considering climate change.

To avoid this moral hazard, the Federal Government should set as its objective that Federal taxpayer dollars provided to States, communities, businesses, or individuals either preor post-disaster be spent resiliently. The Federal Government should not subsidize new development that is constructed in less than resilient ways or in areas at high risk from climate impacts. The government can make immediate progress in this area by focusing on three issues that have been touched upon already—the creation and enforcement of resilient building codes, the provision of accurate risk assessments to inform land use decisions, and the provision of technical assistance to decisionmakers. Thank you very much.

[The statement of Ms. Hill follows:]

TESTIMONY OF THE HON. ALICE C. HILL, SENIOR FELLOW FOR CLIMATE CHANGE POLICY, COUNCIL ON FOREIGN RELATIONS, BEFORE THE U.S. HOUSE OF REPRESENTATIVES, SELECT COMMITTEE ON THE CLIMATE CRISIS, "CREATING A CLIMATE RESILIENT AMERICA: REDUCING RISKS AND COSTS" NOVEMBER 20, 2019

Thank you, Chairwoman Castor, Ranking Member Graves, and Members of the Committee for inviting me to testify today. It is an honor to appear before you and I look forward to answering any questions you may have.

Natural disasters are on the rise. According to the Fourth National Climate Assessment, climate change has already brought more extreme weather and will continue to bring greater extremes in the foreseeable future. The nation will experience a range of climate impacts, including more intense storms, bigger wildfires, and greater temperature and precipitation extremes in the coming decades. Sea level rise has accelerated since the 1990s and will continue to do so in the years ahead. The costs of weather and climate-related disasters are also rising. Between 1980 and 2018, the United States suffered 254 weather and climate-related disasters carrying a price tag of over $1 billion each, according to the National Oceanic and Atmospheric Administration (NOAA). The total cost of these events is more than $1.7 trillion dollars. From 1980 to 2013, the nation averaged 6.3 such billion-dollar events per year. For the years from 2013 to 2018, however, the annual average leapt to 12.6 events. In 2019, the United States has already experienced ten weather and climate-related disasters over $1 billion each, not even counting the wildfires in California. This year is also the fifth consecutive year in which the total number of events has reached ten or more. These figures support the finding of the Fourth National Climate Assessment that the nation's efforts to prepare for climate change impacts have not yet reached the necessary scale to avoid substantial damage to the economy, environment, and human health.

When communities suffer devastation, Americans respond with generosity. The Congress has increasingly authorized supplemental appropriations to provide relief to local and state governments overwhelmed by disasters, to small businesses and individuals who have suffered losses, and to repair damaged federal assets. According to the Government Accountability Office (GAO), between 2007 and 2013, federal appropriations for natural disasters increased 46 percent as compared to the previous six years. In just the last three years, supplemental appropriations for disasters has totaled $183 billion. In light of the growing fiscal exposure to the federal government, the GAO has identified climate change as a "high risk" since 2013.

In addition to greater climate-driven extremes, the increase in damages also stems from decisions made about where and how people build. Those decisions rest almost entirely with the states, and often with local governments. That means that, even though the federal government currently has comparatively little say in state and local choices about land use and construction quality, it frequently picks up the bill for those choices after disaster strikes. The growing propensity of the federal government to absorb the costs of disasters means that state and local governments, developers, and individuals can build in riskier areas and in ways that provide less protection because they believe the federal government will cover the damage when the disaster occurs. In other words, the federal government's growing generosity to victims of disaster creates a "moral hazard": communities and people place themselves at greater risk knowing that federal taxpayers will bail them out. In the face of accelerating climate change, the federal government must reduce the incentives for people to settle in at-risk areas and to build in risky ways.

Take, for example, California's recent experience with wildfire. Ten of the most destructive fires in the state of California have occurred since 2015 and the state is currently fighting a vicious wildfire season. California holds the dubious record for having more buildings destroyed by wildfire than all other states combined. A recent study has estimated it has more than 2.7 million people and 1.1. million homes located in areas at very high risk of fire. In 2008, California enacted a strict new building code designed to reduce fire risk. During the devastating 2018 Camp Fire, only 18% of the 21,100 homes built to older versions of the code survived the fire. Those built to the new, stricter code performed much better, but only 50% of those homes survived.[4]

With only half of the homes built to the latest code withstanding climate-fueled wildfires, California cannot assume that its building codes will keep people and property safe. California faces worsening wildfire risk. Its own climate assessment estimates that climate change will likely expand burn areas 77% by 2100. Just days after the Camp Fire, however, Los Angeles County approved a new 19,000 home development in an area that the state had determined is already at "high" or "very high" fire risk based on past risk and without consideration of the increased fire risk from climate change. If those houses should burn under the worsening

[4] The Council on Foreign Relations takes no institutional positions on policy issues and has no affiliation with the U.S. government. All statements of fact and expressions of opinion contained herein are the sole responsibility of the author.

conditions brought by climate change, it could be the federal government that pays, not those who made the decision to build in an area at high risk.

To avoid this moral hazard, the federal government should set as its objective that federal taxpayer dollars provided to states, communities, businesses, or individuals—either pre-or post-disaster—be spent resiliently. The federal government should not subsidize new development that is constructed in less than resilient ways or in areas at high risk from climate impacts. The government can make immediate progress in these areas by focusing on three issues: (1) creation and enforcement of resilient building codes, (2) provision of accurate risk assessments to inform landuse decisions, and (3) provision of technical assistance to decision-makers.

Resilient Building Codes

Building codes reduce risk of damage. According to the 2019 Edition of the ISO National Building Code Assessment Report, Florida's implementation of a statewide windstorm building code reduced losses by approximately 72 percent. Effective building codes also "have a strong positive effect on disaster preparation and resilience," as the recently released National Mitigation Strategy noted. A 2018 study by the National Institute of Building Sciences (NIBS) found that designing buildings to meet the latest model building codes yields a national benefit of $11 for every $1 invested. In light of the proven value of building codes in reducing damage, the federal government must insist on compliance with resilient building codes where federal taxpayer money underwrites construction.

The United States does not have a national building code. Instead, non-governmental organizations, develop model codes and revise them periodically. The decision as to whether and which model building codes to adopt rests with states and, in many instances, local jurisdictions. Despite the case for strong building codes, however, the Federal Emergency Management Administration estimates that only 32 percent of disaster-prone jurisdictions have adopted disaster-resistant building codes. That means that close to 70 percent of disaster-prone jurisdictions are at greater risk of damage, damage for which the federal government will often be called upon to pay. The federal government must require state and local jurisdictions to use the latest model building codes when building with federal money, either preor post-disaster. Enforcing requirements to adopt and comply with the most recent model codes would save the federal Treasury substantial funds and spare local communities unnecessary damage.

Notably, virtually none of the current model codes, however, yet incorporate consideration of the future risk of climate change. Rather, they rely on historical risk to determine the extremes which structures should withstand. The nation urgently needs model codes that account for the future risk from climate change impacts over the life of a structure. Estimates for when the building code organizations will have developed such codes range to as long as decade. The nation cannot afford to wait that long. As those model codes are in the process of development, the federal government should create its own climate-resilient code for two of the most damaging impacts from climate change—wildfire and flood. Those codes would apply to construction where federal taxpayer dollars are used.

The federal government already has experience with creating climate-resilient codes. Because no model code for climate-exacerbated flooding exists in the United States, the Obama administration, in the wake of Superstorm Sandy and based on the recommendation of the Hurricane Sandy Rebuilding Task Force, developed the first national flood standard, the Federal Flood Risk Management Standard (FFRMS). The FFRMS required that where federal taxpayer money was used to build structures in or near flood plains, those structures had to be elevated to avoid future climate-exacerbated flooding. Ten days before Hurricane Harvey poured approximately four feet of rain on the Houston area causing record flooding, President Trump rescinded the order creating the FFRMS. With the FFRMS, the federal government proved it was capable of producing such standards quickly and efficiently. The nation needs to take advantage of that capacity.

Risk Assessment in Land-Use Decisions

Just as building codes reduce risk, restricting new development in at-risk areas reduces risk. The federal government should not use taxpayer dollars to support new development in high risk areas. Doing so contributes to the moral hazard that those making the decisions to allow development in risky areas do not bear the risk of those decisions.

There is abundant evidence that people are moving into high risk areas. People like to live along our coasts—40 percent of Americans now live in a coastal county— and alongside rivers and streams. These areas face growing flooding risks from climate change, be it more intense storms bringing higher storm surge, sea-level rise, or extreme precipitation, or all of the above. For example, in the state of New Jersey, developers have built almost three times as much housing in coastal flood areas as in less risky areas since 2009. Yet the seas are rising. An estimated 360,000 homes are at risk of permanent inundation by 2050 and 3.4 million homes nationwide could face regular inundation by 2100. People also like to live near forests and grasslands, or what is known as the Wildland-Urban-Interface (WUI), areas. Although living in the WUI often carries a higher fire risk, it is the fastest growing land-use type in the United States. For example, the state of California has more people and property located in the WUI than all the other states combined. It has close to 4.5 million homes and 11 million people in the WUI. Yet, according to the state's own climate assessment, the areas burned by wildfire are expected to grow by 77% by 2100.

The federal government has already acted, albeit in a limited way, to restrict federal subsidies for development in risky areas. In the 1970s and 1980s, Congress realized that the federal government's support of development on high-risk coastal barriers did not make economic sense. The Coastal Barrier Resources Act of 1982 (CBRA) makes certain areas ineligible for federal investments and financial assistance which would encourage development in designated areas. This means that those who want to live and invest in those areas bear the full cost of development and rebuilding after a disaster. According to one Department of Interior Study from 2002, the estimated savings to the federal government would reach almost $1.3 billion from 1983 to 2010. This legislation could serve as a model for restricting support for new development in other at-risk area in the United States. At a minimum, the federal government should not provide financial support for new development in at-risk areas.

To help communities better understand their risks and to guide decisions by the federal government as to which areas are safe to invest in, the federal government needs to provide comprehensive risk maps that include future risk from climate change. That means an immediate concerted effort to create flood and wildfire maps that are updated on a regular basis. Having clear assessments of risk readily available should improve local decision-making and better protect federal investments. Where areas are at high risk from climate impacts like wildfire and flooding, the federal government should restrict its investment in new development in those areas and post-disaster assistance. Where states have already invested in mapping, the federal government can adopt those maps where appropriate.

Technical Assistance

The federal government has enormous amounts of data and information regarding climate change risk. Yet those resources are not often easily understood or even accessible to local decision-makers on the ground. As one part-time mayor of a small town in Alabama, which faces risks of costal erosion from sea-level rise and more intense hurricanes, lamented in 2014, ''I don't have a big planning staff, grant writers, or any resources. So how can I even know the size of the threats we are facing—and what can I do to protect the people of my town?'' This mayor is not alone. Communities across the nation need help deciding how best to prepare for climate impacts. Doing so has the potential to save enormous amounts of money. According to a recently updated study conducted by NIBS, investment in risk mitigation can save an average of $6 in damage for every $1 spent in risk reduction.

The federal government urgently needs to increase its technical assistance to local decision-makers. In 2015, the GAO concluded that the federal government's network of climate data remains so disjointed that "decision-makers are vastly underserved." Although decision-making tools and databases rest on numerous federal government websites, it is hard to imagine how busy local officials can make sense of them without guidance as to their merits and applicability. Similarly, the federal government supports various information hubs, including NOAA, the Department of Agriculture, the Department of the Interior, and the Federal Emergency Management Administration. This approach serves various constituencies but fails to provide a customercentric approach. Those that wish to take advantage of the information must wade through the differing formats, locations, and approaches that each individual agency has chosen to pursue. As the GAO recently noted, because of this uncoordinated approach, "federal, states, local, and private sector decision-makers may be unaware that climate information exists or may be unable to use what is available."

As the National Mitigation Strategy recommends, the government "should support nonfederal partners by providing guidance, useable tools, and resources." The lack of readily available authoritative and actionable information has meant that in many locations and settings, adaptation efforts are stalling. Making climate information easy to obtain and understood would accelerate the updating of codes, the revising of zoning maps, improve engineering and architectural design, and speed revision of cost/benefit analysis. In the absence of current federal leadership in this area, attempts have been made by other entities to fill the void, including a civilsociety-based network for assessing, sharing, and supporting applications of climate science called Science for Climate Action Network (SCAN) (for which I serve as an advisor). However, these efforts alone cannot possibly address the increasing demands for actionable information from across the nation. This should be a core function of the federal government.

One immediate step toward accomplishing this goal is to develop a system for providing technical assistance. Such assistance could help guide state and local governments, businesses, and individuals, through the maze of federal programs and information centers already available. Such a system could also aid identification of ways to combine funding sources and navigate differing program requirements. Assisting decision-makers with on-the-ground choices will save not only them, but also the federal government, from substantial damage and leave the nation safer. Federal assistance in this area can yield substantial savings in post-disaster recovery costs if better decisions about where and how to build are made pre-disaster.

In the longer term, the federal government needs to develop comprehensive climate services to support local planning and investment.

Ms. CASTOR. Thank you very much.

Mr. Berginnis, you are recognized for 5 minutes.

Statement of Chad Berginnis

Mr. BERGINNIS. Thank you, Chair Castor, Ranking Member Graves, and the members of the committee. On behalf of the Association of State Floodplain Managers, thank you for the opportunity to testify. Our written testimony details 40 recommendations for you to consider regarding policy changes and priorities to make a more climate resilient America. For the balance of my time, I will weave those recommendations together and highlight some of them by telling the story of Pecan Acres, Louisiana, which has been in the press recently and compelled me to tie a lot of these things to that story. I first became aware of Pecan Acres by reading an article a couple years ago where I read about the plight of an elderly African American homeowner who couldn't afford to maintain flood insurance coverage as required when they get disaster assistance. The tradeoff would be to go without medication. She was on a fixed income and living in a very flood-prone area. In fact, Pecan Acres developed in 1968 as a community of 40 homes in a very flood-prone area having 17 floods in 30 years, and this was before there were any NFIP maps. So would Pecan Acres have been developed had flood risks been known?

In fact, there are thousands of these neighborhoods from Florida to California to Illinois and everywhere in between, and yet we still don't have a complete picture of the flood risk either present or in the future. Among our recommendations related to data is that we need to get job done mapping the Nation, we need to have a regular program to update our rainfall frequency data, and we need to pass commonsense legislation, like the Digital Coast Act, to empower local decision makers.

Back to Pecan Acres. Flood after flood and seemingly no program could help. Flooding misery, the loss of hope. Structural projects from the Corps of Engineers are likely not cost-effective for such a small and low-dollar-value area. FEMA is only available after a Federal disaster declaration and not after local flood events exacerbated by more extreme rainfall events due to climate change. People are stuck. There is a moral hazard. Those least able to withstand disasters are the ones squarely in harm's way. Today there is lots of programs—the Corps, FEMA, NRCS, even the Federal Highway Administration needs to make sure that we have more resilient roads, but they don't necessarily work well together, and they are definitely not timely.

In actuality, we don't even have a good idea of disaster costs, nor do we do a good job of investigating disasters like we do things like aircraft accidents. Congress has authorized interagency groups, like the Federal Interagency Floodplain Management Task Force, but it has few resources to operate, and we haven't had a unified national program for floodplain management report since 1995, so it is hard to put emerging threats, like urban flooding, in context.

Finally, the last administration's effort to have forward-looking solutions for Federal agencies and the things they fund was repealed in 2017. In other words, there is a lot of room for improvement in aligning Federal programs and resilience goals.

Now back to Pecan Acres. A plan started to come together to use two programs, maybe not traditionally thought of, an NRCS Emergency Watershed Protection Program and Community Development Block Grant to acquire, demolish, relocate, and rebuild households out of harm's way. This raises a couple important points when it comes to adaptation and mitigation.

First, we do not emphasize avoidance enough, whether initially when subdivisions are being built or in considering how we deal with the future threat of flooding. Second, our current approaches need to be improved through strength and resilience standards, accounting for social factors and environmental justice, not just benefit costs and creative ways to help, whether it be through mitigation tax credits or programs like the Department of Defense innovation readiness training program or leveraging new flood proofing technologies, such as those tested to the ANSI 2510 standards. Luckily for the folks in Pecan Acres, there is now hope. When the project is completed, the area will be reverted back to wetlands, harnessing the power of green infrastructure to reduce flood risk. While the project is not complete, it is under way and can serve as an example for other projects. You see there are thousands of these situations across America, and as sure as I sit here today, all of the communities and neighborhoods under present and future threat of flooding will not make it. We are already behind in our planning. Complete community adaptation is measured in decades, and it is imperative that we generate forward-looking actionable data, align programs, create new approaches, and come to grips that there will be places too hazardous to occupy and do all of this in such a way that doesn't leave the most vulnerable behind.

Thank you for the chance to testify, and I will be happy to answer any questions.

[The statement of Mr. Berginnis follows:]

TESTIMONY OF CHAD BERGINNIS, EXECUTIVE DIRECTOR, ASSOCIATION OF STATE FLOODPLAIN MANAGERS, BEFORE THE U.S. HOUSE OF REPRESENTATIVES, SELECT COMMITTEE ON THE CLIMATE CRISIS, "CREATING A CLIMATE RESILIENT AMERICA: REDUCING RISKS AND COSTS" NOVEMBER 20, 2019

Introduction

The Association of State Floodplain Managers (ASFPM) appreciates the opportunity to share our views on adapting to climate change, and being a more resilient nation in the face of this new future condition.

The ASFPM and its 37 Chapters represent over 19,000 state and local officials as well as other professionals engaged in all aspects of floodplain management and flood hazard mitigation including management of local floodplain ordinances, flood risk mapping, engineering, planning, community development, hydrology, forecasting, emergency response, water resources development and flood insurance. All ASFPM members are concerned with reducing our nation's flood-related losses. For more information on the Association, its 14 policy committees and 37 State Chapters, our website is: www.floods.org.

Our Nation's Flood Risk Is Increasing Dramatically

Floods are the nation's most frequent and costliest hazard. Every year the costs to taxpayers continue to increase. ASFPM estimates that in the 1990's average annual flood losses were about $5.6 billion. This increased to an average annual flood loss of $10 billion in the 2000s and in this decade will likely double again to around $20 billion per year.

Climate change is manifesting itself in several ways as it relates to flood risk. But the two primary ways are sea level rise and more intense storms. For the former, the impact of rising sea levels depends on the pace and magnitude of the change— two factors about which there is great uncertainty. For instance, a 2016 study updated the estimates on the amount of ice melting in Antarctica concluded that the increase in sea level may be twice the level that was previously estimated. And, an additional source of uncertainty is the willingness and ability of the world's nations to change the trajectory of climate change. The success of agreements like the Paris Climate Conference and future agreements hold the potential to mitigate some of the projected impacts of climate change.

In inland areas, all across the country, local officials are observing more intense rainfall events. And this is showing up in the data[5] too. Warming conditions mean more water vapor in the air. When rain-triggering conditions are favorable more saturated air leads to heavier precipitation. One public works official from Arkansas recently noted "It was easier when we could plan for and put in stormwater infrastructure that can handle 1–2 inches of rain each hour, but now we are seeing events where you might get four inches of rain in a half hour, I am not sure how we are going to handle that." Recent research[6] by Climate Central reinforces this observation showing an upward trend with more days with 1", 2" or 3" or even more rainfall events.

To meet today's challenges planning for future flooding conditions, while there are promising approaches, overall we are already behind as a nation. ASFPM would like to discuss several areas where improvement is needed. We will address:

- Data, Analysis and Information
- Federal Agency Programs and Policies
- Adaptation and Hazard Mitigation.

---

5 https://www.climate.gov/news-features/featured-images/prepare-more-downpours-heavy-rain-has-increased-across-most-united-0.

6 https://www.climatecentral.org/news/report-pouring-it-on-climate-change-intensifies-heavy-rain-events.

Data, Analysis and Information

If we do not have robust systems in place to provide updated and anticipated hydrologic data, track disaster losses, analyze events, and provide sufficient resources going to research and development, we will simply never get ahead of new development in flood risk areas.

One trend that we are seeing all over the country is that rain events are getting more intense. To compound matters, our nation tends to use outdated hydrology which only further underestimates the risk. The National Oceanic and Atmospheric Administration's (NOAA) National Weather Service (NWS) has been updating precipitation frequency estimates for various parts of the United States and affiliated territories. Updated precipitation frequency estimates, accompanied by additional relevant information, are published as NOAA Atlas 14 and are available for download from the Precipitation Frequency Data Server (PFDS). It is these data that are used in everything from hydrologic modeling for producing flood maps to thousands of design decisions every day for development and redevelopment in our communities throughout the nation. However, NOAA has neither the budget nor mandate to provide this in a timely way. In fact, a note in NOAA's most recent progress report which was through March 2019 indicated that ''No funding is available to extend NOAA Atlas 14 coverage to the remaining five northwestern states: ID, MT, OR, WA, WY in Volume 12.''[7]

Consider the new Atlas 14 data[8] for Texas that came out last fall. That data basically determined that the 100-year rainfall amounts for Houston is now about a 25year event. In Austin, the previous 100-year rainfall amount is now about a 50-year event. As one of ASFPM's Texas members put it, "pretty much all of the flood maps in the state of Texas are now outdated." *And this particular Atlas 14 update was not even looking at the future; rather it is updating 40–50 year old data that was developed in the 1960s*

---

[7] Hydrometeorological Design Studies Center Progress Report for Period OCTOBER 2018 to MARCH 2019, page 4. https://www.nws.noaa.gov/oh/hdsc/current-projects/progress/201904_HDSC_PR.pdf.

[8] https://www.noaa.gov/media-release/noaa-updates-texas-rainfall-frequency-values.

and 1970s. ASFPM is supportive of current NOAA efforts to test the feasibility of incorporating future climate projections into precipitation frequency analysis examining the inclusion of such data into future Atlas 14 updates.

- NOAA should be given the mandate and full budget to update our nation's rainfall frequency information at least every 10 years and this update must include future climate projections into precipitation frequency analysis.

Stream and tidal gages are the stethoscopes of our hydrologic network. Ask any local official about a critical data need and most will say that there needs to be more streamgages. Yet funding for even those deemed critical by the federal government is in short supply. For example, the Federal Priority Streamgages (FPS) Network (previously known as the National Streamflow Information Program) was conceived in 1999 to be a core, federally funded network. The original network design included 4,300 then active, previously discontinued, or proposed new gages that were strategically positioned across the country to address long-term Federal information needs (such as supporting NWS flood forecasts, or interstate and international compacts and decrees). At present (2018), more than 4,700 locations meet the criteria for inclusion in the FPS network, but only about 3,600 FPS are active because of funding limitations. These active FPS are supported through a combination of Federal and partner funding—less than one-quarter are fully funded by the United States Geologic Survey.

- Congress should fully fund our critical national stream gauge and tidal gauge networks.

Another critical piece of data that influences thousands of development decisions every day as it relates to flood resilience are FEMA's flood maps. Since 2012, FEMA has been mandated to not only provide flood maps for the entire nation[9] but also provide future conditions flood risk information. Why future conditions? A 2013 study prepared for FEMA estimated that the 100-year floodplain area would increase by 45% nationally by the end of this century. Yet, little progress has been made on either since that time. In the continental United States, we have 3.5 million miles of streams rivers and coastlines. Yet, FEMA has only mapped floodplains on 1.2 million miles of them. While the FEMA Technical Mapping Advisory Council (TMAC), a congressionally-authorized advisory committee is helping FEMA oversee the nation's flood mapping program, completed the Future Conditions Risk Assessment and Modeling[10] report in December 2015, it appears little has been done and we have yet to have these data appear on FEMA flood maps or in the data provided to communities. ASFPM has previously prepared a programmatic cost estimate for implementing FEMA's National Flood Mapping Program which includes both of the aforementioned mandates, concluding it will cost between $4.5 billion and $7.5 billion to "get the job done" in initially mapping the nation. We note and appreciate Chairwoman Waters' and Castor's efforts to highlight this issue by circulating and signing a dear colleague letter in March 2016 calling for an infusion of funding over five years to complete the job of mapping the nation.

- Congress should provide adequate funding to finish the job of providing flood mapping for the nation, to include future conditions mapping, in a short (5to 10-year) timeframe.

Today's flood maps are based on models that incorporate hydrologic information and topographic information. Good progress has been made on high quality topographic information for the nation through the United States Geological Survey (USGS) 3D Elevation Program (3DEP). These high quality topographic data inform critical decisions that are made across

---

[9] Statute requires FEMA to provide 100 and 500 year flood data in developed areas and areas that have the potential for future development. Since the owner of a tract of land has the legal right to develop, this mandate can be construed as needing data for the entire nation.

[10] https://www.fema.gov/media-library-data/1454954261186c348aa9b1768298c9eb66f84366f836e/TMAC_2015_Future_Conditions_Risk_Assessment_and_Modeling_Report.pdf.

the nation every day ranging from immediate safety of life, property and long-term planning for infrastructure projects. Currently at 60% complete, the goal of 3DEP is to complete the acquisition of nationwide high resolution elevation data by 2023.

- Congress should ensure that the USGS 3DEP program is fully funded to provide nationwide high quality topographic information for the entire nation.

Even if good flood data is developed, there are some policy hurdles preventing it from being publically available. For example, the US Army Corps of Engineers (Corps) new policy[11] on Emergency Action Plans (EAPs) requires several types of flood inundation mapping (EC 1110–2–6074). This policy standardizes inundation mapping and establishes inundation mapping requirements for dams and levees. In theory, having inundation mapping available to the public can help avoid debacles like those we witnessed around Barker and Addicks Reservoirs post-Harvey when thousands of homes in inundation areas of those structures were impacted. Had local land use planners, property owners and others been aware of these risks, steps could have been taken to reduce that risk. However, the new EAP policy includes the following statement: *EAP maps are considered sensitive data and must be marked "For Official Use Only" according to AR 380–5 and DoDM 5200.01*. In other words, inundation maps associated with EAPs are not publically available. Why would we be withholding this vital information on flood risk from property buyers and owners?

The 2016 TMAC report National Flood Mapping Program[12] Review, identified a legacy DHS policy through its Security Classification Guide for the Protection of Critical Infrastructure and Key Resources, which listed dam failure inundation maps as "For Official Use Only." However,

---

[11] https://www.publications.usace.army.mil/Portals/76/Publications/EngineerCirculars/EC_1110 2-6074.pdf?ver=2018-01-22-100438-250.

[12] https://www.fema.gov/media-library-data/1474555532007-c063547f6f48026feb68c4bcfc4116 9d/TMAC_2016_National_Flood_Mapping_Program_Review_Updated.pdf.

this policy conflicts the National Flood Mapping Program requirements that such areas be provided on Flood Insurance Rate Maps and on publically-available databases such as NLD and NID. As noted in the report, a Virginia law passed in 2008 essentially requires that all inundation mapping developed for state-regulated dams be made available to communities and the public. This has now been implemented for a decade without issues and state officials there believe in supporting wider public availability of these data. More recently, when speaking to agency officials, there has been a mistaken belief that this issue had been dealt with. It is clear to ASFPM that it has not and the unwillingness of agencies to act on it demands congressional intervention.

- Congress should mandate that any flood risk data, including all dam/ levee inundation mapping, developed by the federal government and/or associated with any federal program be made publically available.

As a nation, we neither have the system to effectively track disaster losses nor analyze them comprehensively in order to learn lessons that we can apply to future resiliency efforts.

Despite the frequency and expenses of natural disasters, there exists no system in either the public or private sector for consistently compiling information about their economic impacts. Any data collection effort should focus on the losses as a result of natural disasters, or negative economic impacts. The loss from a disaster is a broader concept than its cost, a term that conventionally refers only to the losses that are reimbursed by insurance companies and governments through disaster relief. A National Academies of Sciences report[13] on this topic made several good recommendations that ASFPM supports including recommendations for also tracking disaster payouts incurred by federal agencies to improve tracking federal disaster spending—not only to

[13] https://www.nap.edu/read/6425/chapter/1.

individuals and businesses but also to communities and even spending on repairing federal facilities such as levees or Dept of Defense facilities.

- One agency of the federal government should be made responsible for compiling a comprehensive database containing the losses of natural disasters and disaster spending.[14]

One vital, yet inexpensive, doable step is to adopt the culture of learning from mistakes that we show in other contexts. Consider aircraft accidents. After each crash, we don't gather around the crash site, mourn, confine our blame to the hapless pilots, and solemnly promise to ''rebuild the aircraft just as before.'' The investigation is handled by a standing, independent federal agency, the National Transportation Safety Board (NTSB). Investigators immediately report to the crash scene. They analyze flight recorders and other data to understand the actions of pilots and crew in response to the emergency conditions, but they do not stop there. They go on to consider possible design flaws in the air frame, errors in equipment manufacture, irregularities or shortcomings in airline inspection and maintenance, air traffic control procedures, the prevailing weather—in short, all aspects of aviation that might have any bearing on the incident. Moreover—and this is not so generally appreciated—the NTSB coordinates and leads the team, but the team includes experts from all the stakeholders—the airframe manufacturer, the airline, the FAA, etc. Finally, though NTSB findings and recommendations do not carry the force of law, stakeholders ignore them at their peril. The result? A safety record that has steadily improved over the years with very few aircraft deaths resulting. Something similar is needed with respect to analysis and evaluation of the entire range of all major natural disasters.[15]

---

[14] The National Academies of Science Report identifies the Bureau of Economic Analysis (BEA) within the U.S. Department of Commerce, in consultation with FEMA and other federal agencies involved in natural disaster preparedness, response, and mitigation activities, as best suited for this purpose.

[15] For a more in-depth discussion on this concept (and from where this text was excerpted), please see the June 2006 essay by Gina Eosco and Bill Hooke in the Bulletin of the American Meteorological Society. https://journals.ametsoc.org/doi/pdf/10.1175/BAMS-87-6-751.

- Congress and the administration ought to work together to explore the establishment of a standing National Disaster Reduction Board (NDRB), to analyze and report on disasters. Each report would provide opportunities and incentives for communities and businesses, and state and federal governments, as well as policy makers like Congress to learn from mistakes and make ongoing adjustments to decisions and policies.

For the past decade, a novel approach to data management, tool development and data dissemination has been piloted at NOAA through the Digital Coast Partnership. Developed and maintained by NOAA, hundreds of organizations and federal, state, and local agencies have contributed to this curated collection of high-quality authoritative data and tools focused on coastal and ocean issues. ''More than Just Data'' is the slogan of the Digital Coast because data alone is not enough, especially when users of that data do not know how it can be used, or what steps to take to get information they need. Digital Coast tools and training help users turn data into powerful information that continues to increase the coastal knowledge of our nation. For example, one of the most popular tools being used by practitioners today on the Digital Coast website[16] is the Sea Level Rise viewer. ASFPM was a founding member of the partnership and strongly believes that to better understand the future flooding risk in coastal areas and manage that risk, programs like Digital Coast will be vital.

- Congress should pass the Digital Coast Act.

The House bill (HR 2189) was reported favorably out of committee in September, last week the Senate bill was reported out of committee.

---

[16] https://coast.noaa.gov/digitalcoast/.

Federal Agency/Programs and Policies

While there are numerous programs and federal agencies that address the threat of flooding and floodplain management, most do not take into consideration the future flood condition that will be exacerbated by climate change.

In 1975, Congress established the Federal Interagency Floodplain Management Task Force (FIFMTF). Its purpose was to carry out the responsibility of the President to prepare for the Congress proposals necessary for a Unified National Program for Floodplain Management. For more than 40 years, some form of an interagency group has worked to better understand the appropriate roles of local, state and federal governments in reducing flood losses, the interactions between human actions and natural systems in the floodplain environment and to make recommendations to reduce the loss of life and property caused by floods. Also, the task force is useful to identify and address policy or programmatic conflicts among federal agencies that may be resulting in poor floodplain management decisions. The main report of the FIFMTF, a Unified National Program for Floodplain Management was first written in 1979, then updated in 1986 and last updated[17] in 1995. Unfortunately, the report hasn't been updated in almost 25 years while the threats resulting from flooding have exploded. Not only is research showing significant social impacts of flooding, new flooding types like urban flooding are emerging.

In 2012, ASFPM analyzed more than 130 federal programs that had some impact on the use and development of floodplains. At the time, our evaluation also looked at climate adaptation as it pertained to these programs which, for most was either non-existent or just beginning to be explored.

---

[17] https://www.fema.gov/media-library/assets/documents/18472.

Administration/Congress

- Update the Unified National Program for Flood risk management to define the appropriate role of local, state, tribal and federal governments in managing flood risk including future impacts of climate change and the emerging threat of urban flooding.
- Convene a task force of national economic experts to review and make recommendations for possible changes regarding economic planning and evaluation for flood-related projects; including application of discount rates, treatment of residual risks, land valuation, lost opportunity costs, valuation of green infrastructure and ecosystem services and functions, future conditions and other considerations regarding structural and nonstructural approaches in evaluating flood risk reduction and flood hazard mitigation projects.
- Codify Executive Order 13653—the Federal Government, as well stakeholders, must manage climate change risks with deliberate preparation, cooperation, and coordination in order to effectively improve climate preparedness and resilience.
- Codify an effective federal flood standard when using fed funds to build/rebuild that would address ordinary and critical facilities (e.g., hospitals, water supply, etc) and include consideration of future conditions and a requirement for agencies to consider natural infrastructure alternatives.

Interagency Coordination

- Adequately resource the Federal Interagency Floodplain Management Task Force (FIFM–TF) to better equip it to undertake its role in interagency coordination.
- Direct the Federal Interagency Floodplain Management Task Force to determine how the federal agencies can collaborate on data,

programs and funding to reduce flood risk and flooding costs for taxpayers at all levels.

- Ensure that projects conducted or funded by federal agencies are reflected on FEMA floodplain maps in a timely manner.

Department of Defense/U.S. Army Corps of Engineers

- Investigate the resiliency based standards passed in the McCain Defense act last year that in essence requirc DOD facilities to be looking to higher standards and future climate standards. Determine to what extent DOD has developed rules, is implementing, and is complying with the Congressional mandate and intent.
- Require the development and transition federal planning principles to a National Economic Resilience and Sustainability standard instead of the current National Economic Development (NED) standard to explicitly incorporate the values of multiple ecosystem services, including the non-market public values provided by the nation's floodplains, and future climate conditions.
- Require a minimum design standard of the 500-year flood or PMF level protection for levees protecting urban areas.
- Cease federal taxpayer funding of beach nourishment if benefits are primarily for recreation. Those who benefit should pay for this temporary benefit. The entire beach nourishment policy should be revisited in light of a changing climate and sea level rise. In particular, the cost share for these projects should reduce federal taxpayer costs share to no more than 50%.

Federal Highway Administration

- Improve sharing post disaster highway data and best practices to improve resilient reconstruction of non-federal/state highways.

Develop guidelines to assist local highway departments to help them in reconstruction following flooding.

Housing and Urban Development

- Permanently authorize the CDBG–DR program to avoid HUD having to write rules after every disaster supplemental to streamline the rebuilding process.
- FEMA Public Assistance and HUD CDBG and other disaster funding should require net zero carbon emissions for project eligibility.

Federal Emergency Management Agency

- Under NFIP, consider extending the mandatory purchase requirement for flood insurance to all areas. At a minimum, it should be expanded to other known flood hazard areas such as residual risk areas, urban flooding areas, .2% chance (500-year) floodplain, etc.
- Under NFIP, flood maps must include future flood conditions for NFIP regulation as directed by Congress. Added future flood layers for 2040, 2060 and 2100 projections can be in the digital data for community use for planning or risk commination or other community needs.
- Emphasize the most basic but most important resilience strategy for the NIFP: "avoidance." We should not invest any mitigation money in a community unless they first adopt higher standards that prevent adding any structures or assets within high risk areas. Simply put, we have to stop the vulnerabilities from increasing first and only then start chipping away from what we can then call legacy vulnerabilities.

- Require all Class 7 and better in the NFIP's Community Rating System communities to consider and plan for anticipated climate change in their floodplain management plans. Class 1 communities should prepare maps and regulations using best available data to address the impacts of changing climate for the next 100 years.
- Establish a national flood risk disclosure law to all potential buyers know the past history and future flood risk potential of all properties.
- Require utility companies (eligible for PA) to analyze the full range of mitigation options and account for current and future flood risk in planning, design, construction and reconstruction of facilities. Future federal assistance should be prohibited unless such requirements have been adequately incorporated.

Adaptation and Hazard Mitigation

Community and individual adaptation to climate change will not be quick nor easy. Any community facing flood risk often is also facing a multi-decadal timeframe to reduce that risk enough that they will be resilient in the face of current and future flood threats. Property owners facing increased sea level rise have a very real prospect of their property value plummeting to nothing—for the single asset that, for most Americans, is their most valuable.[18] To say we have an adaptation problem in this country is vastly understating the issue and delay will only add hundreds of billions of dollars in estimated flood related damages that will already likely occur due to climate change.

In some communities, coastal in particular, it is not going to be feasible to stay along the coast given the risks from sea level rise and resources available to adapt. We will need to take proactive strategies and

[18] An insightful 2016 article by Freddie Mac's Economic and Housing Research Group (Life's a Beach), discusses potential impacts of climate change that may be unavoidable when it comes to flooding and concludes that they will likely be greater in total impact than the housing crisis and Great Recession. http://www.freddiemac.com/research/insight/20160426_lifes_a_beach.page.

provide technical assistance to help communities make more informed decisions on when to rebuild more smartly vs when it would be time to start phasing in relocation. Developing innovative assistance programs like the Digital Coast to support the evaluation process, decision making and potential infrastructure/community moves would be important to advance progress. Below are some recommendations:

- Develop national hazard resilience standards for the location, design, construction, and reconstruction of all public infrastructure and buildings that consider: alternative locations, future conditions, green or nature based options, mitigation and a No Adverse Impact approach. These standards should then become a condition of federal funding.
- Minimize use of federal taxpayer dollars to rebuild in areas we know have greatly increasing flood risk.
- Incentivize mitigation through changes to the tax code like a mitigation tax credit.

Flood mitigation actions like buyouts and relocations in particular, will be effective in adapting to climate change, especially in communities where the flood hazard area becomes too difficult for continued occupation. However, our current programs for buyouts and relocations have several issues which make them too time consuming and complex to be done in the manner that they need to be implemented. Congress should examine the buyout and relocation programs that are offered by multiple agencies (FEMA, HUD, USACE, NRCS) to ensure that they are streamlined to the maximum extent possible and also support area wide or community wide buyouts/relocations. In fact, largely due to the complexity of such a project and the inability of federal programs to work together, we rarely see these options used on a large scale. An exception to this is the community relocation project[19] of Newtok, Alaska where both FEMA Hazard Mitigation Grant Program and Pre-Disaster Mitigation grant funds

---

[19] https://content.govdelivery.com/accounts/USDHSFEMA/bulletins/2692581?reqfrom=share.

are being used, as well support from the Bureau of Indian Affairs and the U.S. military through the innovative DoD Innovative Readiness Training (IRT) program.

- Place priority on buyouts and relocation as a way to adapt to climate change.
- Ensure buyout programs/projects pair buyout assistance with the development of affordable housing in less flood-vulnerable areas.
- Fund research on evidence based buyout practices and dissemination of the results to practitioners. Require the FIFM–TF or other task force to examine the hurdles to community wide or neighborhood buyouts / relocations, with a focus on federal programs working together.
- Explore a more widespread usage of the DoD Innovative Readiness Training Program for flood mitigation projects, especially community/ neighborhood relocations.
- Permanently authorize the Community Development Block Grant—Disaster Recovery program.

Congress needs to address the lack of buyout program for flood-prone land in rural areas. Such areas are often those places next to be developed and it would be significantly less costly to acquire either permanent easements or the properties outright then to do so after development occurs. In many areas of the country more floodplain land is needed to safely accommodate flood water through leveed stretches of river. While urban buyouts will improve public safety and reduce property damage, portions of floodplain that are currently protected from flooding by levees must be utilized to convey floodwaters away from towns and critical infrastructure. At the moment, no comprehensive program for land acquisition to improve flood management in rural areas exists. Agencies like the USDA, the Army Corps, and FEMA have various limitations and restrictions on acquisition or easements that make land acquisition a primary barrier to floodplain reconnection projects.

One example would be to improve the USDA Emergency Watershed Protection Floodplain Easement Program (EWPP–FEP). Floodplain easements allow for restoration of natural and beneficial functions of floodplains on land that has been damaged by flooding and allows for floodplains to be utilized to safely convey flood water on undeveloped land. However, this emergency funded program is only activated when infrastructure damages reach a critical threshold to automatically trigger a Stafford Act Federal Emergency Declaration, or if Congress declares easement funding to be available through an emergency appropriation. Unfortunately, both avenues are difficult to achieve. First, the critical infrastructure damage thresholds are almost impossible to reach in many rural counties. Second, if flood damage is localized it can be hard to garnish the requisite national attention needed for an emergency appropriation bill. This can leave rural landowners with unfarmable, flood-prone land following a flood disaster.

- The EWPP–FEP program should be reformed to allow for the release of funding based on more locally based flood damage thresholds or set up as a non-disaster easement program.

While buyouts and relocations are good long-term solutions, there must also be options available in the short to medium term. One approach in the short and medium term timeframe is to use the latest floodproofing technologies. There is an incredible amount of innovation occurring right now as new technologies are coming online to help solve flooding problems. However, are these technologies as good as promised? For buyers, one way to achieve some certainty is to ensure that the product has met the ANSI 2510 standard. ASFPM, in partnership with FM Approvals, assisted with the creation of the 2510 standard over a decade ago. The standard applies to floodproofing technologies such as perimeter barriers, opening barriers, flood mitigation pumps, backflow valves, and now sealants and glazing systems. ASFPM, in partnership with FM Approvals and the Corps oversees the National Flood Barrier Testing and Certification program where products that have been tested and certified to the 2510 standard can be found on the website: https://nationalfloodbarrier.org/.[20] ASFPM is encouraging communities to adopt the 2510 standard and also incorporate it into the nation's building codes.

- Require federal agencies who purchase and use flood fighting products and federal grant programs that authorize the use of such products ensure such products are 2510 certified and are used in floods that meet that certification.

Social and Housing Considerations

More and more, there is a nexus of issues surrounding disaster losses, climate change, social issues (i.e., the effects on low/moderate income (LMI) populations and social justice) and housing. The moral issue is this: How/why do we put those who have the most to lose during a flood in harm's way through our housing, zoning, infrastructure, and other policies? Unfortunately, this is exactly what federal policy does. For example, the Department of Housing and Urban Development does not have a universal policy against paying for housing in flood prone areas. At the same time, we recognize that much of the nation's affordable housing stock was built before climate change was well understood, and many affordable housing options are at risk of flooding. Thus, under current policies, the extreme shortage of affordable housing for low income families is squarely at loggerheads with the realities of flood risk. According to a recent study,[21] nationwide about 450,000 government subsidized households are in mapped floodplain.[22] Therefore, if HUD were to withdraw support from all properties in the floodplain it would create a new crisis of homelessness creating a whole new set of problems.

---

[20] https://nationalfloodbarrier.org/.

[21] https://furmancenter.org/research/publication/housing-in-the-us-floodplains.

[22] As stated earlier in this testimony this is likely a gross underestimate of the housing units at risk given that the nation's flood maps are not yet completed and they do not account for future conditions.

- HUD should examine its housing programs and create innovative mechanisms (i.e., targeted flood mitigation programs for existing at-risk affordable housing units) to incentivize communities, housing authorities, and landlords to undertake mitigation actions with a long-term goal of substantially reducing or eliminating flood risk.
- Incentivize the location of new affordable housing to ensure that it is in flood risk free areas.

The Association of State Floodplain Managers appreciates this opportunity to share our observations and recommendations with this Committee. For any questions, please contact Chad Berginnis, ASFPM Executive Director at cberginnis@floods.org (608 828–3000), or Larry Larson, ASFPM Sr. Policy Advisor at larry@floods.org (608–828–3000).

Ms. CASTOR. Well, thank you all very much. All right. So we have a climate emergency. We do not have time to wait and what we are—what you all—your message to this committee is that we have got to figure out some ways to incentivize local communities and states to do better, to integrate the climate risk into all of the decisions they make. Before I came to Congress, I was a county commissioner, and it was difficult enough, notwithstanding climate factors, just to have sound land use decisions and planning decisions.

How do you all recommend that we from the Federal level here build in the incentives that local communities need or disincentives to ensure that they are in every decision they make in their capital improvement programs, land use decisions, that climate risk is a part of that decisionmaking? Mr. Fugate, if you would start.

Mr. FUGATE. It may not even be Congress that does that. The institutional lenders, the raters, such as Moody's, are already putting local governments on notice that bond ratings may be affected about the resiliency of their tax base and that State and local governments that are not taking steps to address the resiliency of that tax base, the ability to service that long-term debt, may result in higher bonding cost. As you know as a county commissioner, your bond rating is a golden standard for the ability to operate. So I think the private sector's actually further along than we are. Their big question is they don't even know how big the exposure is, but the fact they are asking the question, I think, is the first step. I think the other part of that is, the incentives that we currently have in the disaster programs are such as you have heard from everybody, it is a disincentive for governments to change what they are doing. I think we need to look at things like disaster deductible in the Stafford Act. I think we need to look at quit writing policies for new construction, and I think we need to make investments where we are able to make investments at standards not based upon past weather and past history, but what potentially is going to be the impact.

Ms. CASTOR. Thank you. Judge Hill.

Ms. HILL. I agree with everything that Craig has said. One of the important things is for—we need to make sure that we are informing Americans about their risk. This is a task that I think the Federal Government is particularly well suited to with its strong strength in science and research, to provide the mapping of two American citizens to have better knowledge. And then, once that mapping is in place, which must include the future risk from climate change, we must create carrots and sticks for local communities in terms of signaling that Federal taxpayer dollar will not support new development in areas that are at extreme risk. We will define what that risk level is, but no Federal taxpayer money should support new development in those areas. And then, of course, we will have to address the substantial development that has already occurred, as Chad has indicated, in areas that were not known to be a risk, but we must address their needs as well.

One of the important things we need to do as a nation is develop model building codes that reflect the risk of climate change. We have currently no such model building codes available. Estimates have run that it will take a decade. The Federal Government could provide incentives to our model building code organizations to speed up, to accelerate the development of codes that will help individuals protect themselves as well as communities. Without those codes, we are at risk of building things that will be destroyed in the next flood.

We also need to require that communities who want Federal support follow those codes. And, finally, as has been touched upon, we need to make sure that we are providing the assistance on the ground to people like yourself, Chairwoman Castor, who are trying to make important decisions but may not have all the information needed or all the——

Ms. CASTOR. And there are plenty of communities that are on the front lines that don't have the wherewithal, that don't have even a planner on staff, those kind of issues.

So, Mr. Berginnis, what does this mean for Federal flood standards, and how would communities deal with that?

Mr. BERGINNIS. Well, I think the starting point has to be data and one of the roles—an appropriate role of the Federal Government is to provide good, actionable data and decisionmaking tools, and so, whether it is finishing the job of mapping the nation, whether it is—and this is probably the most egregious issue, is that we don't even have a mandate nor budget to update our rainfall frequency information.

So, in Texas in 2018, they got a new Atlas 14, which essentially basically said, what was the 100-year rainfall event in Houston is now the 25-year event. In Austin, it is a 50-year event, and one of our member's comment is, like, okay, all the flood maps in Texas are now invalid. We are basing thousands of decisions on data that is old and not usable. And so, we need to have that investment of data and one thing that this Congress could do right now, there is the Digital Coast Act that is pending in Congress to work on that data need.

Once we have data in terms of standards, we still have—we still have not done a good job of tying our disaster assistance programs, for example, to the requirement of having standards. What would happen if public assistance, which is by far the biggest amount of disaster assistance we have, was conditioned on the fact that you have latest codes?

I guarantee you almost every single community in the Nation would have the latest codes pronto because you can't turn down that much assistance, but we don't tie appropriate disincentives to our programs.

Ms. CASTOR. Great. Thank you very much. Mr. Carter, you are recognized for 5 minutes.

Mr. CARTER. Thank you, Madam Chair. I appreciate the opportunity, and thank all of you for being here. I appreciate it very much. Do we want to pause?

[Disturbance in hearing room.]

Ms. CASTOR. Thank you very much. You heard me already in my opening remarks say we have a climate emergency, so will you— thank you very much for participating in this great democratic process. Can we proceed to get to the solutions? But thank you very much. Mr. Carter, you are recognized—thank you.

[Disturbance in hearing room.]

Ms. CASTOR. Thank you very much. Mr. Carter, you are recognized for 5 minutes.

Mr. CARTER. Thank you, Madam Chair, and, again, thank all of you for being here.

I want to ask you something that I continue to remind my colleagues of. Do you know what the number one forestry state in the nation is?

Mr. FUGATE. No, sir. We grow trees in the South like people grow corn in the Midwest.

Mr. CARTER. Absolutely. We are blessed in the State of Georgia. We have the number one commercially available timberland, number one in annual timber harvest volume; number one in the exporter of pulp, paper, and paper board mill products; number one exporter of wood fuel; and the number one exporter of wood pellets. And I say that to tell you that, in the First Congressional District, we have got some of the most competitive timberland in the Nation, and I am very proud of that, and I am very proud of the resiliency that is offered through our sustainable forest. And I want to point that out because Mr. Fugate, would you agree that working forests can help to bolster the resiliency of local landscapes, both through reducing soil erosion and improving water quality as well as a number of other things?

Mr. FUGATE. Yes, sir.

In the South, we grow to silviculture. It is not a natural process. It grows trees for production. However, they are not as resilient as we would like. The timber industry in the Florida panhandle and what we call the 850 has been devastated and will take decades to recover. Southwest Georgia saw the same thing. During extreme droughts, all that timber becomes major wildfire areas, and as we continue to see our communities build into the interface, we have developed wildfire risk on the East Coast that may not be as great as the West Coast, but is certainly changing the dynamics of that area.

But that crop and that ability to plant trees obviously does a lot to put land that may not otherwise be usable into productive use, absorb carbon, and help build resilient economies, but it is not totally resilient to the impacts of climate.

Mr. CARTER. Absolutely. And I understand that, but that brings me to a point that I want to bring up. I have got some legislation. It is H.R. 1444, the Forest Recovery Act. Currently, under current law if a working forest is struck by catastrophic loss, as you point out, often happens, hurricanes, wildfires, whatever it may be, 70 percent of timber farmers must simply eat the cost of that. What my bill does is to say that, if they are to repurpose their land that they could get a tax deduction for that and that this would help us. The key there is repurposing it. What we don't want to see happen is for them to lose the land or to turn it to some other use. We want to see them continue to have it to be forest land, and that is what my bill does and what it encourages because resiliency is extremely important and our forests are extremely important to that as well.

Let me move on to talk about community resiliency because it is important today to understand—I also have the honor and privilege of representing the entire coast of Georgia, including 110 miles of pristine coastline, and our coast has been hit by these natural disasters that you mention. Three years in a row, we had hurricanes and this year we just barely missed one with Dorian, but we did miss it, but we still—this is something that impacts us very much. Mr. Fugate, how urgent do you believe it is that we bolster our communities and make them more resilient to withstand these types of weather events?

Mr. FUGATE. I think we can't talk about it anymore; we need to do it. My mama's from Screven. I used to go down——

Mr. CARTER. Your mama's from Screven.

Mr. FUGATE. Yeah. We used to go down to Jekyll Island. I grew up on that part of the world. And what we know, both across all of the Gulf Coast areas and the Atlantic Coast is, we built communities for the past. And when people talk about resiliency, we don't have a good measure. So I would like to introduce a measure because I think this would go right in line of what you are looking at. We need to start looking at the resiliency of the tax base of these communities because we are talking about infrastructure and other things, but what it ultimately comes down to and what Moody's and others are concerned about is, what is the financial risk that communities have and what are they doing to offset that risk.

And this goes back to where and how we build means that tax base will be there after disaster. We are seeing in the 850 panhandle right now Jackson County, Marianna, and other places that their property values have decreased and are not coming back. We saw this in Hurricane Andrew in Homestead City when the Air Force base closed. We are seeing this in Paradise from the wildfires in California. Those communities don't have a tax base. And as the chair will tell you, when you are a local official and your tax base is decreasing at the same time demand for services are increasing, you go into a death spiral and you can't recover. So I think we need to talk about resiliency of tax base and use that as the first nationwide measure of where our vulnerabilities are and where we need to be investing to ensure that communities have resilient tax bases. Mr. CARTER. And that is a great point. And probably the most important point there is just how we should be working with local communities as well, and that is extremely important. Madam Chair, I am out of time, but thank you very much.

And I yield back.

Ms. CASTOR. Thank you, Mr. Carter. Thank you. [Disturbance in hearing room.]

Ms. CASTOR. We are—thank you very much. We are working on climate solutions. We don't have time to waste and that is why I am going to go to Mr. Levin for 5 minutes. Thank you very much. Mr. Levin, you are recognized for 5 minutes. Please. I don't want to have to ask you to leave. I really don't, so would you let Mr.

Levin ask—no, no, no. We get to ask the questions. We will be happy to——

[Disturbance in hearing room.]

Ms. CASTOR. Yeah. Thank you very much. True. We have had those hearings. Okay. Mr. Levin.

Mr. LEVIN. Thank you, Chair Castor. I want to echo your earlier comments. I believe we are in a climate emergency, and I am on the resolution—I am on the resolution stated as such. There is a— all right, please.

[Disturbance in hearing room.]

Ms. CASTOR. Please, I don't want to have to ask you all to leave. Can you please listen to the testimony, and then help us develop these solutions? Mr. Levin.

Mr. LEVIN. In California, we are experiencing a climate emergency firsthand in real time, and it is not some scientific theory. It is not some hypothetical. We are seeing it every single day with extreme wildfires, year-round wildfires that are causing unprecedented problems in our state. Ms. Hill, you said it well. We see planned power blackouts affecting millions of residents, and I am trying, as best as I can along with our California delegation, understand how we best deal with it from the Federal level, so I would like to ask both Ms. Hill and Mr. Fugate. Mr. Fugate, thank you for your service during the Obama administration. What sorts of wildfire mitigation projects have been the most effective in your experience at reducing the risks and costs of wildfires?

Mr. FUGATE. Actually, two programs that are not Federal. The first one was California coming up with better building codes. We know particularly in the Camp Fire that the fire started in embers. The roof materials were a big factor in flame spread, and so I think California's taken that, but we have a built infrastructure of homes that aren't there yet. Another one was a program that was instigated by the National Fire Protection Association called Fire Wise. Steps we can take right now to reduce vulnerability to homes such as landscaping, just managing debris around homes, ladder fuels, and things that can reduce the impacts of wildfires, and then you as Congress, after the disaster 2017, reauthorized or made changes to the Stafford Act. For the first time, we have permanent authorization to provide mitigation dollars under fire management grants that can again go back, but we also have to look at the fire risk as a two-stage threat. It is the fire that does the initial devastation. Then it is the flood risk and the mud flows afterwards that will be caused by all the scar burn.

So, again, I think what California took was a first step. I think there are programs like Fire Wise that can give us immediate tools to help communities and homeowners reduce their risks, but I also think we need to look long term, as was pointed out by Judge Hill. Where we build and how we build is the future, but how do we take care of the people that are already in the interface.

Mr. LEVIN. So, at the Federal level, how can we supplement some of those best practices.

Mr. Fugate. Again, I think you are providing mitigation dollars which will allow communities to now start doing things other than structural mitigation. We have learned a lot about how we can control fuels around our yards, the vegetation management. I know this is a controversial issue, and I am from Florida where we do silviculture. We do a lot of controlled burning to manage vegetation. We learned after 1998 that if we don't do control burns, we end up in the interface with uncontrollable wildfires. That is not a very easy discussion to have in California, but I think it is one that has to be looked at, is fuel reduction during the extreme wet periods so that when you go into the drought periods, there is less fuel available, it is managed when it can be managed, and it will help reduce as we have seen in other parts of the world as well as Florida. It can reduce the severity of the wildfires, but it is not a silver bullet. There is a lot of things you have to bring together to start seeing that curve bend.

Mr. LEVIN. Ms. Hill, briefly, if you can answer the same question because I would like to move on to some other things, but what are best practices you have seen in terms of mitigating wildfires? What can we do on the Federal level to be most supportive?

Ms. HILL. Thank you. I do agree that it is a difficult area, but prescribed burns are proven to reduce the fuel, and that would keep the fires to be less hazardous. I also think that we need to look at supplementing or helping communities look at their risk as a community because it is not—if one house catches fire, it may well be that other houses catch fire. And if in a community you have an individual living in a home who does not have the means, for example, to replace a wood shake roof, which is like piling kindling on top of your roof, according to firefighters, we need to help that community come together to make the entire community safe as a result of fire.

We are just at the beginning of this. I don't think there is deep understanding of how these fires interact with the built environment, and we need to fuel more money to have research in fire dynamics so we understand better how we can safely build and live in a wildland-urban interface.

Mr. LEVIN. Mr. Berginnis, I want to follow on and talk about the effects of floods in the areas that have experienced wildfires. Could you talk about whether Federal programs really appreciate the risks of floods that follow the wildfires, or what should we be doing differently—in particular, you know, providing adequate support to the communities that are working to address both the risk of the fires and the floods.

Mr. BERGINNIS. Sure. I think the Federal Government is beginning to figure that out. The passage of the DRRA in 2018, of course, now triggers hazard mitigation with any FMAG declaration, fire declaration, and it is kind of two sides of the coin, right? After a wildfire, your next biggest threat, of course, is going to be flooding and dealing with that. And also Craig had talked about Fire Wise as a program, and, again, I kind of go back to the concept: Congress can implement these things in terms of a technical or financial incentive or take away disincentives for not doing things. And so, again, a program like Fire Wise is a voluntary program right now much like in the Flood Insurance Program, the community rating system is, but at what point in our communities that are facing extreme risks, at what point do we try to normalize those programs or make them required or mandatory so that they are incorporating the latest risk reduction approaches as part of community's business. And so I think far too communities are on Fire Wise much like—or far fewer communities are in Fire Wise that need to be and similar with the CRS. So whether the Congress does that by linking that as a requirement in order to get financial assistance or whether it incentivizes it through some sort of additional financial or technical assistance are two options.

Mr. LEVIN. I am over time, but I thank you all very much for your answers. Look forward to working with you.

I yield back.

Ms. CASTOR. Ms. Bonamici, you are recognized for 5 minutes. Ms. BONAMICI. Thank you so much, Chair Castor.

Thank you to our witnesses for being here today. I just want to mention briefly that the group exercising their First Amendment rights has left the room at present, but I want to say: We hear them. We understand. They are really concerned, and they have a lot of anxiety about what they see for their future for this planet. So I am a cosponsor of my Oregon colleague Representative Blumenauer's bill to declare a national climate emergency. We need to send a message that we must take action, so I just wanted to put that on the record that there is that bill, but we really need to take this seriously. And I understand, although I don't agree with the disruption, I do agree with them raising their voices in a peaceful way to get the message across.

So, according to the fourth national climate assessment, by 2025, if we don't address our aging and deteriorating infrastructure, it is going to cost up to $3.9 trillion, close to $4 trillion to repair and replace it. More than 60,000 miles of our roads and bridges are already experiencing extreme storms and hurricanes that are costing billions to repair. We know that sea levels are rising. Could be one to four feet by the end of the century, devastating many coastal communities including—I represent the north coast of Oregon. The frequency and depth and extent of tidal flooding is expected to continue to increase in the future with coastal storms and today's infrastructure and building standards simply do not take those future trends into account, and that is why we are here today to talk about that.

Current levels of infrastructure investment in the United States are not enough to respond to the threats of the climate crisis, and I don't know if this was brought up before I arrived, but my other Oregon colleague, Mr. DeFazio, who chairs the Transportation and Infrastructure Committee is talking about making those infrastructure investments in a sustainable way.

Ms. Hill, in your testimony you discuss how the Federal Government's resources and data on climate change risks is not often easily understood or even accessible to local decision makers on the ground. The example you provided of the mayor in Alabama that sort of echoes the concerns that I often here from communities across Northwest Oregon, especially those smaller communities where, as Chair Castor was saying, they might not have planning staffs and people to do this work to sort out and decipher all the Federal information.

So how can Congress better support a synthesis of existing Federal research and assessments and identify gaps to help especially our local governments?

Ms. HILL. We need to have a customer-focused approach. After Hurricane Sandy, the Hurricane Sandy Rebuilding Task Force worked hard to make our programs comprehensible to those who needed to access them. Right now, we have datacenters across the United States set up by various agencies—FEMA, HUD, agriculture. Those all service their particular constituencies and fulfill their missions, but for that part-time mayor or that planner who has got a busy schedule, it is very difficult to make sense of our programs. They sometimes have conflicting requirements. We need to streamline this with a view to making it accessible to those who need Federal support, and we also need to make sure that every single one of our programs is screened for climate resilience to make sure that we are not inadvertently supporting development that is not resilient.

I would say, in the area of infrastructure as well. We need to look at our cost-benefit analysis. Our cost-benefit analysis is not permitting us to make the types of investments in resilience that we need to have for very long-lived structures.

Ms. BONAMICI. Right. It requires some long-term thinking. Ms. HILL. Yes.

Ms. BONAMICI. You suggested providing better technical assistance. I absolutely agree that that be done in a way that is accessible to those, especially those front line communities. I hope we can all work together on that.

Mr. Fugate, did I say your name properly. Mr. FUGATE. Yes, ma'am.

Ms. BONAMICI. In your testimony you referenced a study by the National Institute of Building Sciences that found that investing in mitigation saves society $6 for every dollar invested. In the Pacific Northwest, we don't see climate change as a distant threat. It is our reality. It is important to support those communities as they make those investments in resiliency, but when facing multiple threats, wildfires like my California colleague talked about, sea level rise, flooding, extreme weather, warmer temperatures, it could be challenging for local decisionmakers to determine where to put their money first and how to set priorities. How can we better assist communities in determining which aspects of the built environment are most immediately vulnerable to climate change, and do we need to develop some best practices for use on a regional scale?

Mr. FUGATE. Yes. I know the National Institute of Standards has been working on this. We have a lot of people talking about resiliency. We don't have a measuring stick. I don't think we can wait for perfection. We just need something that we can apply uniformly across the Nation, across multiple risks. That is why I look at tax base. There is a cautionary tale to that, though. As Judge Hill points out, cost-benefit analysis is always looking backwards, but it also favors the highest value property. So for Chair Castor, what we would see if I am planning mitigation projects, all things being equal? I tend to apply mitigation projects to the more affluent areas in my communities because they have the highest tax value, and I tend to underprice the more vulnerable communities because they are not going to save as much tax dollars. So I think we need to be making sure that when we talk about resiliency, it is not for the affluent alone.

Ms. BONAMICI. Thank you. I know I am over time, but if I could briefly follow up on that. We are also, in the Pacific Northwest, overdue for earthquake and tsunami because of the Cascadia subduction zone, and I think of places like Seaside, Oregon, a small town over on the Oregon coast where, for years, they worked to pass a bond measure so they could move their schools out of the tsunami zone, and it took a very long time because it takes people stepping up and saying: I am willing to pay more to make sure that my kids are going to go to a school in a place where they are going to survive an earthquake and tsunami. They finally were able to pass that bond. There was some land donated, and they are opening their new school outside of the tsunami zone, and it was students who really made the case. Some high school students made a video and said that we are starting the clock now, and now, 15 minutes later, we are active, healthy high school students, and we are still on flat land, and we are going to be underwater. So it was really compelling, again, the voice of youth helping to make the case. It is resiliency because of a natural disaster, but certainly analogous to what we are facing with the climate crisis. And thank you for your indulgence.

And I yield back.

Ms. CASTOR. Mr. Graves, you are recognized for 5 minutes.

Mr. GRAVES. Thank you. Thanks again for being here, and I apologize that I was unable to be here for your testimony, but I will finish going through your written testimony. Administrator Fugate, you and I have had the opportunity over the years to talk about this appropriate balance of proactive efforts, and you were a big advocate of predisaster mitigation, and we did the BRIC program last Congress. Could you talk about your thoughts on where and how do you strike that right balance for coming in and making resilient investments versus picking up the pieces after a disaster and based on your extensive experience in disaster management and recovery efforts, how do we do a better job with that?

Mr. FUGATE. I don't think we start on the backs of people who already live in flood-prone areas or other risky areas that are not affluent. I am a big supporter of Risk 2.0 that FEMA's looking at to price risk, but also think it has to be means tested.

Mr. GRAVES. You might be one of the only people who actually knows what it is because everybody else calls it a black box, but please, go ahead.

Mr. FUGATE. But it is the idea that we should be pricing risk closer to what market value puts on it, but in your state, in particular, but throughout much of the country, it isn't coastal property with expensive homes that I am most concerned about with this program. It is the existing homeowners that were built in many cases before we had flood data or is underestimated flood risk that if we priced them out of their homes, I don't think that is good policy. But then you have to have the balancing act that is actually now subsidizing development in coastal areas for affluence that is increasing the exposure and risk, so to me we need to take care of what we got. We need to quit growing the risk, and I think that is where pricing market moving back to the private sector, providing less incentives for the Federal Government to step in every time you are having recurring events and then start making investments, not that we are not going to build, but we are going to build differently in ways that are sustainable against future risk.

Mr. GRAVES. How do you—what you just hit on and again going back to conversations years ago, those are tools that are often controlled by the State or the local governments. How do you integrate those into your overall toolbox and planning efforts?

Mr. FUGATE. I am from local and state government and being up here at the Federal level gave me a different perspective, but I always go back to, our theory is decisions are best made closest to where people live. However, Congress has a big checkbook and people are responsive to the availability of funds or disincentives in those funds to change behavior. And I think that is those levers that Congress has, like the National Flood Insurance Program. It requires local governments to adopt building codes that no other program has because the program says if you want to be in the program, you got to adopt the codes, otherwise, we are not going to provide it. I think those are the tools, Congressman, you have to look at is, what are things that governments at the local and state level would be most responsive to and can we give them—not immediately "you got to change, or you are done"—but can we give them a glide path that says, "We are going to give you a chance to start moving in this direction." But if you don't move in this direction, there is going to be pain.

Mr. GRAVES. What do you do for people in south Louisiana, for example, communities that have been there for hundreds of years sustainably that, all of a sudden, because of changes that they didn't have anything to do with, more water coming down, Gulf of Mexico encroaching on them as a result of river management practices by the Corps, how do you treat them? They didn't have anything to do with the vulnerability they are experiencing today. How do you treat those people? How do we address that from a fair policy perspective?

Mr. FUGATE. I don't know if there is fair, and I don't know if there is an easy answer. And I am afraid we are going to be telling far too many people in your state, my state, and others that we are not going to be able to rebuild back the way we were after the disaster and that some communities are not coming back the way they were that have been there for centuries. We are seeing this in the Florida Gulf Coast.

So we have to look at where we can come back, what we can do differently, but what we cannot come back or doesn't make sense, how do we provide the transition for people to pick up and move on with their lives with some certainty versus the inevitable delays and buyout programs and promises that never materialize, and they are back where they were to get hit again.

Mr. GRAVES. Under our coastal master plan, we did just that. We effectively drew a line in the sand and said: If you are below here, we don't have the resources to protect you for whatever other reason we can't, but we did say, we will help provide assistance to elevate homes or to relocate. And I will tell you, that is a really tough decision because people don't deserve it, and it is not fair, as you have indicated.

Ms. Hill, would you care to comment on that balance issue that I asked the Administrator about?

Ms. IIILL. I fervently agree with his comments. We are going to have to have a glide path for those who are in areas that will no longer be safe to live in. The land will either be too soggy or too at risk of wildfire for them to be there or some areas will be too hot. We haven't looked at all this. It will be an equal opportunity disaster for many in America, and we have to address how do we help them get out of situations, which, as you have pointed out, were not of their making? It simply reflects decisions made on historical risk.

We now know that the climate is changing. Building to the past will not keep us safe. Land use decisions made on past—based on— past extremes will not keep us safe, so what do we do with those that are already there? The one thing that is clear: Let's not add any more to those that are at risk. So that should be a bright line rule going forward just as you have said, but we need to figure out how we will help those who are already at risk find safer ways to live and thrive.

Mr. GRAVES. Thank you.

Madam Chair, in closing, I do want to make note, I heard Mr. Berginnis in his opening statement reference Pecan Acres, and I do want to be very clear that we are the ones who secured the funds through NRCS as well as the FEMA funds that are being used for the relocation. I want to give a shout out to General Honore; our chief of staff, Paul Sawyer; and others that have been working on getting that package put together; and Pat Forbes at the state, but thank you for mentioning.

I yield back.

Ms. CASTOR. Mr. Casten, you are recognized for 5 minutes.

Mr. CASTEN. Thank you, Madam Chair. Got here in perfect timing. Thank you all so much for coming.

I really want to dig into this conversation about flood insurance. I sit on the Financial Services Committee. We have had hearings on this. We all, I think, around this whole room understand the core issues that this program that was created as an insurance program, which is really well suited for things that are expensive, rare, and unpredictable, doesn't work as well when they are only expensive. And I think we also all recognize that there is—there is a fundamental equity problem that, if we charge the market rate for insurance, people aren't going to have insurance, not the people in the low-lying areas who most need it and I don't think we have a choice as human beings about whether or not we look out for those in need when they do get flooded. So we still end up paying; we just pay through a different mechanism.

I appreciated your testimony, Mr. Berginnis, and I am going to put you on the spot for the million-dollar question, because if you solve this question, we can all have fewer hearings in Financial Services. How do we solve that basic equity problem? We can't force people to move to the expensive parts of town if they can't afford to live there. We can't charge a rate that is market rate and expect them to pay, but we do have to look out for our fellow man. How would you like to see us solve that?

Mr. BERGINNIS. So I think it is making changes to multiple programs starting off with the National Flood Insurance Program. I agree with Craig 100 percent, risk rating 2.0 is important from the psychological aspect to communicate to folks what risk is based on what the full risk rate is. Now, of course, as you are well aware, within the NFIP, we have a safety net against that with the rate caps, and I think one of the things that Congress is looking at is changing those, once again, which would in effect lengthen the glide path, and so, for folks that may not be able to afford it, it really buys time to be able to do that. Also, I know, in the House legislation, there is a means tested approach to help those that can't afford the insurance. I am always struck by the name of the NFIP because I actually think it is the worst named program in Federal Government. Because in working with it my whole career, it is four programs in one. It is an insurance program. It is a land use and planning program. It is a hazard mitigation program, and it is a data and mapping program. And I would urge Congress that, as you are thinking about changes into one knob, think how it affects the other pieces, right.

And so, for instance, the whole discussion on private flood, one of the concerns that we have raised is the fact that, you know, you have too much going to private flood. It is the NFIP policies that fund 100 percent of the floodplain management efforts within the program. So we have got to figure out how to do that. But then, outside of the NFIP, so you have those lower income property owners that are having a hard time with paying for the insurance. They are making the decision between the medications, food, and insurance. They are in a really horrible situation, and how can we help mitigate that? We have not in our mitigation programs put enough emphasis on kind of end-to-end resiliency and what I mean by that is that we tend to focus on, okay, you know, what is the safe and what is the most cost-beneficial approach? What we don't focus on is the fact that folks need affordable, safe sanitary housing to be able to go to. And so we need to make sure that we are marrying up, whether it is a mitigation program by FEMA with HUD programs to make sure that we are funding the relocation assistance that they need or else even, I think in the Pecan Acres approach, they are actually funding to build new affordable houses for folks to live in, and I think we have got to look at that end of it. Mr. CASTEN. So I totally agree with you on that. Do you think that that program is sufficient if we don't also have covenants on the redevelopment of the land after we move them off those low lying areas?

Mr. BERGINNIS. No. I don't think it is sufficient from the standpoint either of the resources we have and the one thing that I do have a concern about, and it goes back to the data issue, and even on the example I talked about with Pecan Acres, what is the flood risk in 50 or 100 years on the site that they are going to? So we are making big investments now, but if we don't have the data we are still flying blind. And so we got to have that data too.

Mr. CASTEN. As Mr. Graves started by saying that I was not in a coastal area, hopefully, this will all increase property values in Illinois sixth as people come to an area that is not——

Ms. CASTOR. Flooding is everywhere.

Mr. CASTEN. Yes. Thank you so much. I yield back.

Ms. CASTOR. Thank you. Mrs. Miller, you are recognized for 5 minutes.

Mrs. MILLER. Thank you, Chairman Castor, and thank you— well, he is gone. And thank you all for being here today. In 2016, West Virginia suffered catastrophic floods that resulted in the loss of lives of 23 people. Our communities are still in recovery mode today. We currently have critical infrastructure that needs to be rebuilt, and I want to ensure that, as we move forward, that this process when we are rebuilding, that we have resilience at the top of our mind.

Mr. Fugate, in your testimony you discussed about how after a disaster we must rebuild for our future risk. How can we better engage with our State and local governments on this topic?

Mr. FUGATE. Well, part of it is FEMA, but I remember we went round and round not to rebuild those elementary schools back where they flooded. We got heat from local officials, we got heat from parents, people that lived there all their generations didn't want to move those schools. They would be further away. And we stood our ground, but we also were fighting an internal battle at FEMA that our cost-benefit analysis wasn't supporting the decision to relocate them, just repair them as they were, and I am like: Are you insane? We are going to put elementary schools back where it flash flooded? We were fortunate school was not in session, but what would have happened if those schools had been occupied.

We have to give clearer guidance to FEMA that cost-benefit analysis isn't about an insurance policy looking at how much it saves the taxpayer if we do something different. It is about the function. It is about the life safety. And in many cases, it won't even be the dollar value. It is the societal impacts of not doing something that we need to address.

Mrs. MILLER. While pre-disaster mitigation helps save money after a disaster occurs, the upfront investment can be cost prohibitive for local communities. What can be done to make pre-disaster mitigation more affordable?

Mr. FUGATE. If we are only going to spend money to fix stuff we didn't do right the first time, it ain't going to work. We have to invest in pre-disaster mitigation for built infrastructure, but we also have to encourage local and state governments to adopt the building codes and land use standards to quit making it worse, and that means we have to build to the future.

The other thing is we got to do a better job disclosing risk. You know that most states don't require you to disclose prior flood risk? That I can go buy a home that has been flooded, and there is no requirement to disclose it? We need truth in advertising. I think this is something else that is pretty straightforward we could do, but either here or in the flood insurance bill is required, if you are going to participate in the National Flood Insurance Program, you disclose flood risk and flood history before you go to close.

Mrs. MILLER. In certain states, vulnerable or low-income populations usually do live in areas that are more at risk for a natural disaster. How can we empower this population to engage in pre-disaster mitigation on their homes?

Mr. FUGATE. Bring them to the table. I remember a Deputy Secretary said: Nothing about you without you.

And I think we do not engage the public and the communities at risk in these discussions. I think when they make the decisions, we are much more successful. When they look at the information, they look at what the options are, but if you look at most of our programs it is always government to government, and we tend to leave out the communities we are trying to serve.

Mrs. MILLER. Well said.

Ms. Hill, in your testimony, you discussed how local communities lack technical assistance. I know many communities in rural Appalachia that could certainly benefit from such a program. How would you suggest deploying such a program around the United States, particularly in rural communities?

Ms. HILL. We need a cadre of trained personnel who can take the tools and data that currently exist and help communities understand how those materials can assist them in their decisionmaking, but if we only leave our information and our data on the web for some busy person to flip through screens to try to determine what decisions they should make, a community should make, I do not believe we will get to success. We need to have individuals who represent the Federal Government, not necessarily their solely their agency, but are familiar with all the programs that will help communities be stronger and can wade through the different regulations, the different requirements to get them to success.

Right now, our approach is solely based on what is convenient, in my opinion, for the Federal Government. We need to flip this. It is what works for our communities. We need to have invest in their interests in learning what their risks are and how they can do better.

Mrs. MILLER. That is good. From your experience in this field, do you think the public and community leaders adequately know what resiliency means and what role they have in preventing it?

Ms. HILL. No. I don’t think most leaders have the opportunity to learn about climate risk and what it may mean for their community. It is difficult to find accurate information, even based on past risk much less future risk.

This is an urgent issue because we are seeing—and forgive my framing it this way—stupid decisions being taken across the United States now in very expensive investments. There is no way to screen currently to make sure that Federal investments in infrastructure are resilient, and that means that we see many examples that have occurred in recent history with huge dollar price tags attached which will not be resilient to risks that will unfold during the life of the structure.

That seems to me irresponsible use of taxpayer dollars, not to insist that, if we hope the bridge lasts 50 years, we understand what the risks are to that bridge, and build a bridge that can withstand it.

Mrs. MILLER. With so many different types of geography in our country—and I know in my particular area, most people are settled down at the base of mountains, but when you start talking coastal and just—you can't have the same plan everywhere, and so, when you talk about having educated people for X, Y, and Z, you wouldn't send the Z people to the X; you would send the right people to the right district to help educate them.

Ms. HILL. Absolutely. One of the best ways to start is scenariobased planning with the community. Meet with the leaders and stakeholders in the community, have the charts, the maps, the visualization of what the future risk is and work through what is your hazard mitigation plan, what is your future plan, and what is it going to look like after you have the next flood, and we need to make the decisions about where you will rebuild in a safe manner for those who enjoy your community.

Mrs. MILLER. Thank you. I yield back.

Mr. BERGINNIS. May I——

Ms. CASTOR. Go ahead.

Mr. BERGINNIS. If I could add just a little bit to the questions, Congresswoman. I wanted to underscore something that Craig had said in terms of both codes and land use standards and understanding that land use standards are as important as having strong codes. So, for example, when you are laying out and designing new subdivisions—and I worked in Ohio, so next door neighbors to West Virginia in the Appalachian foothills, I worked with those communities, and there is a lot that can be done. One of the things that ASFPM developed is a guide on 60 optional higher standards to make your subdivisions safer and we provide that to States and communities, but the other piece of this that I think is missing that hasn't been talked about yet is State capacity.

The best way we helped our smaller communities, our more impoverished communities in Ohio is, when we have the state capacity at the state level, that knowledgeable cadre of experts in the state floodplain program or in the state mitigation program to help do that. And so, in the National Flood Insurance Program, there is something called the Community Assistance Program, CAP– SSSE and, in fact, Mr. Casten's committee, Financial Services, in the NFIP reform bill is looking to make that a statutory program. In the mitigation side, we don't have anything that is parallel, and what CAP does, it actually helps fund state level floodplain management experts who can in turn help communities. On the mitigation side, when you look at most state mitigation programs, they have maybe one, two, three people in there. They probably need to have ten or 20 people, and I would argue that having an incentive program like CAP–SSSE for mitigation can help get you to the objective that you are seeing.

Mrs. MILLER. Thank you. I hope I get one of those books.

Ms. CASTOR. Great. I have just a few questions before we close out, but everyone will have an opportunity if you all would like to ask some other questions as well. I have been able to go to various communities across the country and ask them what they would need as we develop recommendations for a climate action plan involving adaptation, and to a community, they have said, we need more predisaster mitigation resources to help us with this.

Do you all know what scale are we talking about? What should we be considering as we follow through with that recommendation? Mr. FUGATE. I will give you a number, and you are not going to like. If you look at what the backlog currently nationally of infrastructure projects and you look at mitigation and what it would take for predisaster mitigation to adapt, you are talking a trillion or more dollars. That is why we got to quit growing the risk. We are never going to get caught up. Things such as National Flood Insurance Program, we got to take care of what we got, but we got to quit growing the risk. Just stop writing flood insurance for new construction and let the private sector market manage that.

We need to get stronger building codes and land use planning, and then we need to start looking at—it will not be an equal impact with climate change across the country. And if you look at the identifiable threats we are going to see first, it is the extreme rainfall events, the extreme heat, drought, and wildfires, and sea level rise will not affect every area the same. And this is easier to talk about in this body, but next door in the Senate, they are very much 100 divided by two is how they calculate financial distribution. And as much as we fought this with Homeland Security, we are going to have to make prioritization on our limited funds that we are not going to be able to fix everything in every community, but we are going to have to come up with a bench mark of how do we measure our resilience exposure. Our tax base would be one way, and then make investments to start buying down that risk where one technology exists, the practices occur, and we have good data. Quite honestly, some of this we don't even have yet, and I am not sure making investments there would give us the fastest return.

As the folks in the back said, even if you stop all of your emissions today, we are not going back. The only question is, how much worse is it going to get?

Ms. CASTOR. What else can you all add on the scale of pre-disaster mitigation?

Mr. BERGINNIS. When I think about the scope of just repetitive flood loss properties in the country, we have over 150,000 of those. That means those are the ones that are going to be more vulnerable than the ones that just have flood insurance and really not had the claims. And then you multiply out the value of the home, and you are talking an astronomically huge number. I think we also need to think about the concept of kind of a force multiplier and how can we incentivize States to invest in mitigation programs.

You know, the State of Texas just earlier this month passed a constitutional amendment that sets up State-funded mitigation program basically for flooding. Finally, I think after 3 years of 500year floods,

Texas got it, and they are now beginning to invest some of their own funding, and I think part of the solution is making sure that the State has mitigation—has their own mitigation program available that can help, especially things that are not declared at all.

The other thing I would say is that, even though there is the theory and the desire of predisaster mitigation investment, we have to always remember the psychology of disasters, and I have been to communities, I have tried to sell FEMA's predisaster program, and on a sunny day when nobody is displaced, most of the time, they say: We don't need to do mitigation. We don't want to go through that hassle of elevating.

And so there is always going to be a role for postdisaster because, unfortunately, it takes the disaster to change the mindset that, hey, we actually need to be more resilient as property owners.

Ms. CASTOR. Judge Hill, do you want to add—I have one more question after this.

Ms. HILL. Sure. As has been noted, the scale is really massive, but we must remember: If we don't start mitigating and making these investments now what is the economic cost to the United States as you project this out. And, of course, that is what is behind the Government Accountability Office—of their concern—of placing this on the high risk. One of the bright line rules I think we should adopt immediately is we will not spend in a way that is not resilient going forward. So the limited funds that we have that may not be marked as mitigation—it may be just simply an investment in a grant program—we are not going to spend unless we are assured it is resilient to the future impacts that we expect. Ms. CASTOR. One last question for me, and then Ms. Bonamici can close us out. So a lot of local communities and states have reserves and rainy day funds, but the Federal Government doesn't do that. Instead, we have a natural disaster and then an emergency aid bill that doesn't really sometimes arrive in a community on an emergency basis. Gosh, we have had partisan food fights and even Hurricane Michael, those folks were left out to dry. So, when it comes to the Department of Defense, they have a multibillion dollar overseas contingency fund for contingencies when it comes to defense issues. Isn't it time to be reconfiguring Federal budgeting so that we actually have a rainy day fund of some sorts or catastrophe fund that is front-end loaded and we have those resources ready to deploy?

Mr. FUGATE. Chair Castor, when we went through sequestration we ran out of money for Hurricane Irene and Disaster Relief Fund, we came up with a good plan. I think Jack Lew and Speaker Ryan, and that actually has worked and balanced out, and, except for extreme events, has not required Congress to do immediate funding. In fact, money was there for the Michael response recovery. Unfortunately, the bureaucracy is slowing down getting that money.

But there is one area that I think Congress should act on—the ranking member has been talking about this; he has been looking at how to do this—is, if we are going to use HUD as a major funding for long-term cost of disaster impacts, they need permanent authority. This, every time they get appropriations starting from scratch, delays getting money out for a year or more.

HUD if they are going to be part of this team, they need permanent authority and they need staff. And this should not be we have to create the wheel every time Congress uses that mechanism to address the longer term issues that communities face.

Ms. CASTOR. Any other comments? All right. Ms. Bonamici, you are recognized.

Ms. BONAMICI. Thank you very much. I just want to add another point. Once we get the policy and the funding, we absolutely need the workforce because when we are doing all this work, we have to have the people with the skills out there to actually put it into place. So I have a bipartisan bill, the BUILDS Act, which is about making sure that we have a workforce when we invest in infrastructure. So I came here from a hearing on apprenticeships. It is just a challenge with so many of these, whether it be housing or infrastructure, we need to have the people to do the work so we need to fill that gap as well, the skills gap, so I just wanted to add that.

If we really want to get the work done, we have to have people to do the work, and I thank the chairwoman for allowing me to add that point.

Ms. CASTOR. Well, thank you very much. Thank you to our witnesses. This was an outstanding hearing. I am sorry that there is so much else going on on the Hill today. We couldn't have a full panel of members. And this is a reminder to everyone who is tuned in here that we are seeking policy proposals on the greenhouse gas mitigation side and on the adaptation side. We have a request for proposals, request for information on our website. Please if you have policy proposals for the committee, please send them to us. The deadline is the end of this week. Thank you again for being here. The committee is adjourned.

[Whereupon, at 2:58 p.m., the committee was adjourned.]

United States House of Representatives Select Committee on the Climate Crisis, Hearing on November 20, 2019 "Creating a Climate Resilient America: Reducing Risks and Costs" Questions for the Record, The Hon. W. Craig Fugate Principal, Craig Fugate Consulting LLC Former FEMA Administrator

The Honorable Kathy Castor

1. In your testimony, you emphasized the importance of being able to measure community resilience. How does the tax base of a community affect its resilience? What sorts of climate-related metrics should the federal government use to measure the resilience of a community, institution, or system?

Measuring the Resilience of a community's tax base is a key vital sign of vulnerability to climate change impacts. Global warming immediate and long-term impacts include extreme rainfall events, droughts and increased wildfire risk, more destructive tropical storms, and sea level rise.

Think of the tax base as a vital sign, like a pulse, it will not tell you everything about the community, but like a pulse, lack of one is death. A slow or weak pulse, like a tax base below the cost of running the community's government, will impede the recovery after a disaster or loss of tax base due to sea level rise.

My observations of recent disasters underscore the risk to housing stocks at risk from weather related hazards. From the devastation of wildfires to Paradise, Ca [1] to the impacts of Hurricane Michael on the Florida Panhandle, [2] loss of housing impacts tax base revenues, workforce housing, and the speed of recovery.

Tax Base as a measure of resilience provides the ability to measure multiple hazards and their impacts to a community. Key vulnerabilities to housing, businesses, and critical infrastructure by floods, hurricanes, wildfires, and other natural hazards are based on:

- Hazard Impacts (areas at risk)
- Vulnerability (how and where a community built)
- Restoration time (how long before infrastructure is restored, businesses reopen, and homes are repair or replaced).

Current disaster risk models will need to be updated to look at future climate driven risk.

2. What sorts of wildfire mitigation projects are the most effective at reducing the risks and costs of wildfires? What can the federal government do to encourage communities to undertake wildfire mitigation projects?

In three board areas that can reduce the losses of life and structures in the wildland urban interface or WUI:

- Model building and development codes for development in the WUI [3].
- Fuel reduction [4].

- Homeowner Actions to maintain defensible space around their home [5].

The National Fire Protection Association Firewise program developed guides for homeowners and communities to reduce the risk and impacts of wildfire in the WUI [6].

3. How well do the model codes address the impacts of climate change, including flood and wildfire risks?

We have seen where enhanced building codes have reduce the impacts of major hurricanes in the Florida Keys during Hurricane Irma and the epic wildfires of 2017 and 2018 in California [7].

Current building codes should be seen as a minimum standard, they are often based on past hazard events. This was evident during Hurricane Michael in 2018, the area of impact used a lower wind standard than the rest of Florida based on the past history of hurricanes on the region [8].

With Climate Change increasing the risk of more extreme weather hazards, building codes based on past weather risk history will not build resilient communities. The Institute for Business and Home Safety (IBHS) has developed a standard called Fortified that uses current building codes as the minimum and enhances the code for specific hazards related to Hurricanes and strong winds. From their web site "FORTIFIED is a nationally recognized building method that goes beyond building codes to strengthen residential and commercial buildings against specific natural hazards such as high winds and hurricanes. FORTIFIED standards are based on more than 20 years of scientific research and real-world testing by IBHS" [9].

4. How can Congress redesign federal disaster assistance to move funds more quickly while also assuring more resilient outcomes?
    a) Provide guidance to federal agencies to balance oversight with speed of approval of funding projects under federal review.

b) Provide HUD with authority to permanently authorize the Community Development Block Grant—Disaster Recovery (CDBG–DR) Program. From the GAO ''Without permanent statutory authority and regulations such as those that govern other disaster assistance programs, CDBG–DR appropriations require HUD to customize grant requirements for each disaster in Federal Register notices—a time-consuming process that has delayed the disbursement of funds. In a July 2018 report, the HUD Office of Inspector General found that as of September 2017, HUD used 61 notices to oversee 112 active CDBG–DR grants. Officials from one of the 2017 grantees told us that it was challenging to manage the multiple CDBG–DR grants it has received over the years because of the different rules. CDBG–DR grantees have faced additional challenges such as the need to coordinate the use of CDBG–DR funds with other disaster recovery programs that are initiated at different times and administered by other agencies. HUD officials said that permanently authorizing CDBG–DR would allow HUD to issue permanent regulations for disaster recovery. Permanent statutory authority could help address the challenges grantees face in meeting customized grant requirements for each disaster, such as funding lags, varying requirements, and coordination with multiple programs. The expected increase in the frequency and intensity of extreme weather events underscores the need for a permanent program to address unmet disaster needs.

c) Establish by legislation the requirements of the former Executive Order 13690 and the Federal Flood Risk Management Standard for federal grant program involving the construction of infrastructure and buildings in flood prone areas.

The FFRMS gave agencies the flexibility to select one of three approaches for establishing the flood elevation and hazard area they use in siting, design, and construction. They could:

- Use data and methods informed by best-available, actionable climate science;
- Build two feet above the 100-year (1%-annual-chance) flood elevation for standard projects, and three feet above for critical buildings like hospitals and evacuation centers; or
- Build to the 500-year (0.2%-annual-chance) flood elevation.

References

[1] Wildfire's destruction of California town creates uncharted credit territory https://www.bondbuyer.com/news/california-wildfire-destruction-may-devastate-credits.

[2] With 80 percent destroyed by Hurricane Michael, Mexico Beach struggles to stay livable https://www.wdrb.com/in-depth/with-percent-destroyed-by-hurricane-michael-mexico-beach-struggles-to/article_efab6e5c-33da-11e9-93e5-c75d91998d95.html.

[3] Institute of Business and Home Safety—IBHS supports the use of wildlandurban interface (WUI) codes to help reduce the potential for wildland fires to spread into the built environment. The Wildfire Codes & Standards—State-by-State Reference Guide is a comprehensive assessment of wildfire building codes in each of the 50 states. In addition to the assessment of wildfire-focused codes in each state, the members-only guide includes a glossary of wildfire terminology and IBHS guidance on wildfire-resistant building construction. *Key Findings:* Only four states have WUI specific building codes adopted statewide; Eight states have guidelines or programs to reduce wildfire risk; Where WUI codes exist, enforcement of those codes remains a challenge; https://ibhs.org/wildfire/wildfire-building-codes-and-standards/.

[4] California In a 45-Day Report to Governor Gavin Newsom in response to Executive Order N–05–19, CAL FIRE systematically identified high priority fuels reduction projects and other measures to immediately begin to protect over 200 of California's most wildfire-vulnerable communities and put the state on a path toward longterm wildfire prevention and forest health. https://www.fire.ca.gov/about-us/45-day- report/.

[5] Example from Institute of Business and Home Safety (IBHS) Create Defensible Space. Maintain 3 zones around your structure, collectively called defensible space: Remove dead vegetation; Trim branches overhanging roof; Remove combustible materials in the 0–5 FT zone.

[6] NFPA's Firewise USA® program teaches people how to adapt to living with wildfire and encourages neighbors to work together and take action now to prevent losses. www.firewise.org.

[7] Keys homes, battered but standing, may be a model for reducing damage in Florida https://www.miamiherald.com/news/weather/hurricane/article173408496.html. *California Wildfire Building Codes:* How a building code change could be a pivotal moment in California's wildfire fire https://www.denverpost.com/2019/04/11/california-building-code-wildfires/.

[8] Hurricane Michael exposes building-code weakness in Florida's Panhandle. Until 2007, building-code standards for windstorm resistance were more rigorous in South Florida than in the Panhandle, where major hurricanes have been rare https://therealdeal.com/miami/2018/10/13/hurricane-michael-exposes-building-codeweakness-in-floridas-panhandle/.

[9] Fortified Home—A National Standard for Resilient Construction https://fortifiedhome.org/.

Questions for the Record, The Hon. Alice C. Hill, Senior Fellow for Climate Change Policy Council on Foreign Relations

The Honorable Kathy Castor

1. You discussed in your testimony the role of building codes in community resilience. How well do the model codes address the impacts of climate change, including flood and wildfire risks?

Model building codes in the United States do not yet address the impacts of climate change. The codes tend to rely on historical weather events to account for risk and focus on life/safety rather than building performance. Efforts are underway to create climate-resilient building codes, but it may be years before such model codes exist.

2. How should Congress define resilience and integrate resilience into the laws we are enacting? How can we better prioritize federal investments around measurable resilient outcomes?

The federal government currently uses multiple definitions for the word "resilience." The Disaster Recovery Reform Act of 2018 directed the Federal Emergency Management Administration to issue a rulemaking defining the term. This is a welcome development. Any definition of the term must, however, specifically account for future impacts of climate change. One definition that would accomplish this is: "the capacity to reduce, absorb, and recover from events, including the future impacts of climate change." Without consideration of future risk, resilience efforts will prove futile in the face of worsening climate impacts.

Federal-wide adoption of a common scoring system to measure the resilience of particular investments would assist in the prioritization of those investments. Projects receiving the highest score could receive funding priority.

3. In your testimony, you discussed the role of communities in managing land use and the challenges that communities are facing when the available maps of flood and wildfire risk do not consider climate change and the conditions communities will face in the future. How important are maps that show climate risks into the future to community land use and zoning decisions? How can the Federal government help address these challenges?

Maps can assist in identifying areas at high risk from climate change impacts. Maps can also help state, local, and federal decision-makers, as well as ordinary citizens, better evaluate whether taxpayer dollars should support new or continuing investment in high-risk areas.

The federal government should undertake a nation-wide effort to develop maps that reflect future risk from climate change. A good place to start would be accurate mapping of future risk from flood and wildfire. The federal government should also commit to updating these maps on a routine basis such as every five years.

4. How can Congress increase community insurability and the use by communities of private insurance for assets that would be insured if they were privately owned, such as buildings?

Congress can increase community insurability by providing incentives to states to permit the use of models of future risk to determine insurance pricing. Incentives could take the form of making available additional levels of federal disaster aid, for example, if the state permitted the use of future modelling to help determine pricing. In the absence of insurance pricing that reflects future risk, current pricing practices could result in underpricing insurance and make it less attractive for insurance companies to offer insurance in certain areas. The ability to consider future risk in pricing could ultimately increase the likelihood of insurance companies continuing to offer insurance.

QUESTIONS FOR THE RECORD, CHAD BERGINNIS EXECUTIVE DIRECTOR, ASSOCIATION OF STATE FLOODPLAIN MANAGERS

The Honorable Kathy Castor

1. What is the appropriate role for the federal government in managing flood risk?

The federal government has multiple roles in managing flood risk. Three primary roles are listed below. Perhaps the most important role is the *provision of data and information*. Given the large reach of federal agencies, data is key to effective decision making. That is why ASFPM wholeheartedly supports FEMA's National Flood Mapping Program, the USGS 3DEP program to collect LIDAR (topography) for the nation, USGS and NWS streamgaging, and the Digital Coast Act (which focuses on curating coastal data sets). We testified that a critical gap or need is to have a robust program to update precipitation frequency information. While NOAA produces Atlas 14, there no substantial ongoing funding or mandate to have that data updated every 5–10 years which is what ASFPM thinks is needed.

Another role is *providing leadership* through promoting effective standards, effective program execution, and eliminating perverse incentives to not be flood resilient in federal programs. In 2012, ASFPM research determined that over 150 federal programs had the potential to impact sound floodplain management objectives; however, not all of these are oriented to ensure long term flood resiliency. For example, the tax code casualty loss deduction provides relief to individuals who did not take the step to purchase flood insurance, even if they were required to. An example of a perverse incentive is the authorization of a new flood control project by the USACE as long as it meets National Economic Development objectives. This may mean, a new levee that is built to less than a 100-year standard versus having a minimum standard or requirement for levee resiliency based on public safety (for example the Netherlands uses a 10,000 year standard). ASFPM strongly supported Executive Order 13690 which would have required agencies to consider future flood conditions and adhere to a higher standard than the NFIP is presently. An example of effective program execution is how FEMA does not hold its on-the-ground Federal Coordinating Officers to account for any kind of minimum requirement for building hazard mitigation into public assistance programs or for ensuring hazard mitigation projects get to a certain state of development before making decisions to close or scale back Joint Field Offices. Yet the policy goal of both Congress and FEMA is that both should be implemented fully and expediently.

Finally, a role is to provide incentives and resources to build and local capacity in flood risk management. The Federal government cannot do it all, nor should it. Flood risk management is a joint federal, state, local, individual and private sector responsibility. Too often these days federal program try to be implemented directly at the community level while states are either not included or overlooked. It is much more efficient and effective to build capacity and delegate authority to states.

2. What role should states play in assisting communities to build resilience to climate change? Are all states capable of providing such assistance? If not, how could that capability be improved?

(1) Developing more specific, downscaled data; (2) providing state resources through state level resilience programs, (3) developing state plans and standards for resilience, and (4) providing training and building capacity at the local level. Presently few states are capable of providing this assistance. However this capability could be improved through a mix of carrots (incentives) and sticks (penalties). For example, if the availability of public assistance (by far the largest source of postdisaster aid) was conditioned on the requirement that a community had to have a valid mitigation plan as well as participate in additional resiliency program depending on the hazards they face, they undoubtedly would do it (stick approach—currently if a community doesn't have a hazard mitigation plan

the only penalty is that hazard mitigation funding is unavailable). In fact, many more forms of disaster assistance, including CDBG–DR should be tied to hazard resilience activities.

3. How might states and communities use information about future sealevel rise and flood risk to manage flood risk and reduce future losses?

In a lot of ways. ASFPM and the American Planning Association just released a new report[23] on incorporating flood resilience into capital improvement planning (because infrastructure projects are typically a community's largest investment). Future Sea Level Rise information is beginning to be used in communities along the U.S. Coastline (i.e., New York City, Norfolk VA, State of California) for planning and to implement both land use and building standards. It is being used by state Department of Transportation to do long-range repair/replacement planning.

4. How can federal programs that use a Benefit-Cost Analysis better measure and integrate resilience into those analysis, and prioritize mitigation investments toward more resilient outcomes?

A couple of thoughts. First, is that the discount rate, at least for FEMA's benefitcost analysis needs to be lowered. The effect of the artificially high discount rate in the FEMA BCA methodology limits FEMA's ability to approve mitigation projects that are, in fact cost effective. This is a recommendation recently made in the November 2019 FEMA National Advisory Council report[24] which recommended the discount rate be lowered from 7% to 2–3%. Second, is that most benefit-cost analysis modules do not account for social impacts.

---

[23] https://www.planning.org/publications/report/9192800/.

[24] https://www.fema.gov/media-library-data/1576076713587-c1ae1017f2adb4b836ad3db0c26d3578/November_2019_NAC_Report_final.pdf.

5. How well do the consensus-based model codes address flood risk? What sorts of changes would you recommend to the model codes to address sea-level rise and extreme rain events? Why do you think the model codes haven't integrated the sorts of freeboard requirements that are already in place in so many states and communities?

While consensus based codes have progressed over the years to begin to more proactively address flood risk, they are still far from what is desirable, especially as it comes to resiliency against future flooding conditions. Given that two trends we are seeing in the science is that sea level rise estimates are likely too conservative and slow, a meaningful requirement should reflect the useful life of the type of building which the code applies. For a residential home, this may be well over 100 years. That means we need to begin to build estimates of SLR out to 2125 and beyond. Because model building codes are not land use codes (how to build in a risky area more safely versus whether you should be building there in the first place), the minimum freeboard requirement should be 2 feet and in coastal areas should be 3 feet. Critical facilities should be 3 feet or the 500-year elevation whichever greater. One trend ASFPM is beginning to see is communities adopting the 500-year flood level above the FEMA mapped floodplain as a proxy for future conditions. Another requirement would be to use future conditions floodplains especially for critical facilities.

Model codes haven't integrated the sort of freeboard requirements that are already in place in many states and communities because the process is very hard to get forward thinking ideas approved. And this is typically due the outsized presence of the homebuilding industry in the process.

6. What are the most effective ways to assist communities so that they can build resilience to climate change into their plans and actions?

The most effective way is to have FEMA map future conditions floodplains and include them in the package of information they give communities immediately. While FEMA has been required to include future conditions into their flood map updates as a result of the 2012 reform of the NFIP (after they had been advised by the TMAC as to how to do it), FEMA has yet to implement this future conditions requirement. As was answered in question 1, another aspect of this is that as a nation we must have a mandatory, frequent update of rainfall-frequency information (currently the program to update Atlas 14 does not have consistent funding nor a mandate). Another critical aspect is to help them interpret future conditions scenarios. For example, today, New York City is planning for 6 feet of Sea Level Rise by 2100, while the State of Hawaii is planning for 3.2. Why? This is partially due to the future condition scenario that was picked. Another effective way is to invest in quality datasets and tools that can be used by communities. A great example of this is the Digital Coast initiative by NOAA.[25] In fact, ASFPM supports Congressional passage of the Digital Coast Act (already passed by the House) which would build success on the initiative.

7. How do communities use Hazard Mitigation Plans before and after disasters? What is the relationship of Hazard Mitigation Plans to comprehensive plans, zoning, and building codes?

Hazard mitigation plans are typically developed prior to a disaster but are hardly ever referenced after the disaster or adjusted in the immediate aftermath of a disaster (which would be a best practice). Even worse, there has not been a lot of success integrating these plans into comprehensive plans, zoning, and building codes. There are numerous reasons for this but one of the primary ones is that hazard mitigation plans are often produced by the local emergency manager and comprehensive plans, zoning are led by the local planning department—these silos do exist at the local level too. The FEMA publication[26] Integrating Hazard Mitigation Into Local Planning pages 2–4 and 2–5 is a good summary of where these points of intersection and opportunities for integration exist.

---

[25] https://coast.noaa.gov/digitalcoast/.

[26] https://www.fema.gov/media-library-data/20130726-1908-25045-0016/integrating_hazmit.pdf.

In an ideal world, the hazard mitigation plan, once developed or updated, would be used to then feed into the update of a community comprehensive plan which would, in turn, lead to updated zoning and building code standards. A new approach, the Plan Integration for Resilience Scorecard or PIRS created at Texas A&M University is a hands on, facilitated approach that gets participation from different local government agencies and uses a scorecard to identify points of consistency and points of inconsistency. A link to a webinar on this approach can be found here.[27]

---

[27] https://www.floods.org/ace-files/training/SLIDES_PlanIntegration_PIE_Webinar_10.4.2017.pdf.

INDEX

A

B

C

D

E

F

G

H

I

L

M

N

O

P

R

S